DRAFTING MADE SIMPLE

BY

YONNY SEGEL, M.S.

Assistant Professor,
Bronx Community College

MADE SIMPLE BOOKS
DOUBLEDAY & COMPANY, INC.
GARDEN CITY, NEW YORK

DEDICATED TO TRUDA

ABOUT THIS BOOK

Drafting Made Simple combines the presentation of all the necessary fundamentals of drafting and draftsmanship and a detailed exposition of drafting practices and methods and methods used in specific industries.

The initial portion of this book concentrates upon the hard-core fundamentals of drafting. The reader is taught how to master the various materials and techniques which the draftsman utilizes in practicing his profession. Orthographic projection, dimensioning, the use of scales, auxiliary views, isometric drawing, blueprint reading, architectural drawing—all these and other basic skills are carefully and clearly explained with the visual aid of hundreds of illustrations. You are systematically taken through each type of drawing—step by step—so that you can see the entire process from the initial phase to the completed drawing.

In the latter sections of the book, you are given a clear insight into the drafting methods and practices used in the steel industry, in heating and air conditioning drawings, in electrical and electronics drawings, in piping drawings, in aircraft drafting and in the plastics industry.

An outstanding feature of this volume is the practical problems, with answers provided in the back of the book, which the reader is asked to solve. You will test your ability to draw the three views in orthographic projection; you will try your hand at using scales, at drawing developments and at isometric drawing.

The entire complex procedure of constructing industrial, commercial and residential buildings requires a thorough knowledge of the skills and science of drafting. From the inception of the original plans, through the drawing-board stage and the final construction, the architect, the draftsman and the builder combine their knowledge and talents to produce beautiful, utilitarian and necessary edifices for the continuous progress of man and his civilization.

It is the draftsman's drawings that enable builders to construct modern structures and to develop the entire technology of a scientific culture. In *Drafting Made Simple* you are taught and shown how the draftsman employs his skills, knowledge, and talents to enrich our common life.

ACKNOWLEDGMENTS

I am indebted to the organizations listed below for granting me permission to use illustrations from their books and pamphlets or for granting me permission to use drawings of their manufactured products.

Eugene Dietzgen Co.; Keuffel & Esser Co.; The O. A. Olson Manufacturing Co.; Homemaster Publications, Inc.; Arnold A. Arbeit, A.I.A.; Dunham-Bush, Inc.; The American Society of Heating and Air-Conditioning Engineers; General Electric Company, Transformer Division; Ryan Aeronautical Company; The Braddock Instrument Company for permission to use an illustration of the Braddock-Rowe Lettering Angle.

Figures 265–268 from *Building Trades Blueprint Reading & Sketching—Basic Course*, copyright Delmar Publishers, Inc.

Figures 261, 262, 269, 271, 272 from *Manual for House Framing*, published by the National Lumber Manufacturers Association.

American Technical Society, publishers of *Applied Drawing And Sketching*.

Figures 299, 300, 302–308 from Volume I of *Structural Shop Drafting*, published by the American Institute of Steel Construction, Inc.

Graphical Symbols For Electrical Diagrams-Y. 32-2 1954, published by the Institute of Radio Engineers.

Engineering Manual, published by Bell & Gossett Company.

Aircraft Designers' Data Book by L. E. Neville. Copyright 1950. McGraw-Hill Book Co.

I am especially grateful to the American Society of Mechanical Engineers, 29 West 39th Street, New York 18, New York, publishers of the "American Standards," for the cooperation which the Society gave me in preparing this book and for the permission it granted me with regard to reproducing material from the following American Standards.

B 4.1-1955-Preferred Limits and Fits For Cylindrical Parts

Y 14.1-1957-Size and Format

Y 14.2-1957-Line Conventions, Sectioning and Lettering

Y 14.4-1957-Pictoral Drawing

Y 14.5-1957-Dimensioning and Notes

Y 14.6-1957-Screw Threads

Y 14.11-1958-Plastics

Y 14.15-1960-Schematic and Electrical Diagrams

I would like to express my appreciation to Mr. Jerry Patton, who helped with the illustrations for the earlier chapters, and to Miss Sylvia Feldman, who typed the original manuscript.

Yonny Segel

TABLE OF CONTENTS

TOOLS AND MATERIALS

The draftsman must, of course, have tools with which to work. The following list includes all the basic tools and materials which you will need as you begin your study of drafting.

1. Drawing Board
2. Drawing Paper
3. T Square
4. Triangles
5. Protractor
6. French Curves
7. Pencils
8. Erasers and Erasing Shields
9. Dusting Brush
10. Scales
11. Compasses
12. Dividers
13. Ruling Pens and Ink
14. Templates
15. Masking Tape
16. Lettering Pen
17. Lettering Guide
18. Cheesecloth
19. Razor Blade or Penknife

The beginner will find these items sufficient for use at home, in school, or in the office.

In professional drafting rooms, paper, pencils, erasers, sandpaper blocks, cheesecloth, masking tape, and drafting tables with T squares attached will usually be provided.

The beginner may be confused by the many types and qualities of tools that are available. In general the better-quality tools and materials will help you to do a better and more accurate job. They may also prove to be more economical in the long run.

DRAWING BOARD

Fig. 1 illustrates a drawing board. Drawing boards are made of white pine or basswood and come in various sizes. A board 18″×24″ is adequate for general use. The board should be made of several tongue-and-grooved pieces and should have a cleat at each end to prevent warping. Edges that are opposite each other must be straight and parallel.

If paper sheets 36 inches or larger are to be used, drawing tables will be required. The choice of table will depend on the space available and the amount of money to be spent. The draftsman must work standing up in order to reach all parts of a large drawing sheet. He therefore needs a stool on which to sit when his work does not require him to stand. You will find that most stools are uncomfortable but that the use of a rubber pad will help. You will also find that by tilting your drawing board slightly upward you will be improving your own view.

DRAWING PAPER AND CLOTH

Drawing paper comes in a great variety of sizes, colors, grains, weights, strengths, erasing qualities, rag contents, aging qualities, and transparencies. Small-size sheets, such as those 11″×17″, may be purchased in pad form. A roll of paper available in lengths of from 10 to 50 yards will prove economical for your purposes. White paper is used most frequently in drafting rooms. Cream or green-colored paper, finely-grained to take pencil lines easily, is considered preferable for ordinary use. The United States Patent Office requires that 2-ply Bristol Board be used for trade-mark and patent drawings. You can purchase these Bristol Board sheets in the required size, with the proper border and title already printed on them. The greater the rag content of the paper,

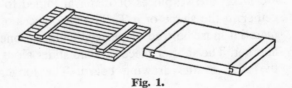

Fig. 1.

the better it is for use with ink. The erasing quality of any paper simply refers to the ability of the paper to withstand ink erasures.

Draftsmen often draw directly on transparent materials such as paper, linen, and glass cloth. This method of direct drawing reduces the cost of reproduction. Water-resistant drawing cloths, specifically treated for pencil or for ink, are manufactured in blue and white tints. Paper also comes lined for graphs, charts, mathematical formulas, business statistics, and for many other uses. The beginner should, therefore, consult a reputable dealer when he wants paper or cloth for any specific purpose.

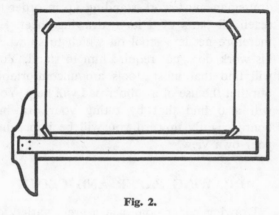

Fig. 2.

Fig. 2 illustrates the proper way to place the paper on the drawing board. The paper should be placed on the board near the bottom of the sheet so that you can easily draw lines with the T square. Since you put the head of the T square on the left side, place the sheet closer to the left side of the board. Use drafting or masking tape on each corner of the paper. Thumb tacks leave holes and also make it difficult to slide the T square. The drafting tape can easily be removed without leaving a mark or tearing the drawing.

T SQUARES

Fig. 3 is a drawing of a **T square.** T squares are made of two pieces of material joined together in the shape of a T. The two pieces of material must be true or perpendicular to one another. The shorter piece or head slides along the edge of the drawing board. The longer

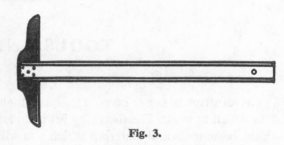

Fig. 3.

piece or blade is used to draw horizontal lines and to support other drawing instruments. T squares are made of wood or metal.

Wooden blades with transparent edges have two specific advantages over those with all wooden blades. The transparency allows you to see lines underneath the blade and also helps prevent blotting. Some T squares have movable heads so that you can draw at any angle.

A straightedge, made from a straight piece of wood or metal, is constructed like a T square except that it has no head. It can be used to draw straight lines in any direction. Large drafting tables are equipped with straightedges which have pulleys and cords so that they can slide on the table. Straightedges may also be attached to small drawing boards with clamps.

TRIANGLES

Fig. 4 is a drawing of three types of **triangles.** Two kinds of triangles are used most frequently. One has 30°, 60°, and 90° angles, while the other has two 45° angles and one 90° angle. Both types come in various sizes. A 10-inch 30°—60° triangle and an 8-inch 45° triangle are adequate for general use. Most triangles are now made of transparent plastic.

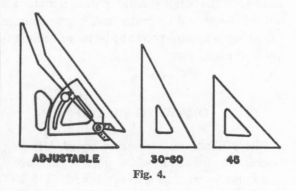

ADJUSTABLE   30-60   45

Fig. 4.

Carbon tetrachloride or soap and water is used to clean most triangles. Some plastics, however, cannot be cleaned with carbon tetrachloride. Triangles made of xylonite can be easily cleaned with carbon tetrachloride or soap and water.

Adjustable triangles, illustrated in Fig. 4, enable the draftsman to draw any size angle. When a drawing requires angles other than 30°, 45°, or 60°, the adjustable triangle is most helpful.

PROTRACTORS

A **protractor,** illustrated in Fig. 5, is used to measure or mark off angles. Protractors are not used to draw angles but simply to measure them. Metal or plastic protractors can be purchased, but the plastic type, which can be washed with carbon tetrachloride, is preferred.

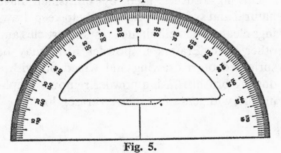

Fig. 5.

FRENCH CURVES

French curves, shown in Fig. 6, are used to draw *odd-shaped* curves. The transparent, plastic French curves shown in the drawing are adequate for most purposes although other shapes are available. For map drawing, curved rulers that can be bent to any curve up to a specific maximum radius are used. The better French curves have transparent ruling edges.

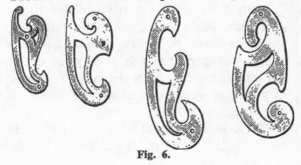

Fig. 6.

PENCILS

Pencils containing leads with varying degrees of hardness are manufactured. Pencils that contain hard lead are usually marked with the letter H or with the letter H and a number which precedes it. For example, a pencil might be marked 2H, 3H, 4H, etc. The higher the number which precedes the H, the harder the lead. Pencils that contain soft leads are usually marked with the letter B or the letter B with a number preceding it. The higher the number which precedes the letter B, the softer the lead. Leads of medium hardness are denoted by the letters HB or F. The draftsman requires a variety of hard pencils. Soft pencils do not give a sharp line and are, therefore, of little value to the draftsman. Pencils marked 6H, 4H, 2H, H, and F will provide enough variety for the beginner. Leads, which may be purchased without the wood casing, are often used in special holders. Leads used in this fashion can be sharpened easily because the wood doesn't have to be removed each time. Besides, some holders may contain a hard lead at one end and a soft lead at the other end. Fig. 7 illustrates a mechanical pencil.

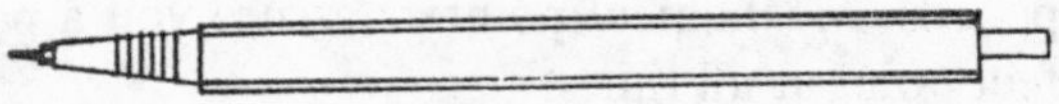

Fig. 7.

PENCIL SHARPENERS

A sandpaper block or a fine steel file is used to sharpen a pencil point. First, expose about ⅜ of an inch of lead, as illustrated in Fig. 8A. Hold the pencil near the middle and rotate it clockwise while moving it toward you along the sandpaper, as demonstrated in Fig. 9. A conically shaped point, like the one pictured in Fig. 8B, can be obtained this way. Most drafting rooms have pencil sharpeners, called draftsman's pencil sharpeners, which give an extra long, rough lead point, as shown in Fig. 8A, to make it easier for sharpening on the sandpaper block.

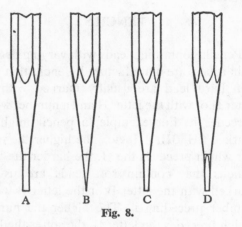

Fig. 8.

Most draftsmen tie the sandpaper block to the drawing table and attach a piece of cheesecloth to the string to wipe the graphite dust off the pencil after sharpening.

Draftsmen sometimes use chisel points rather than conical points. After exposing enough lead, as shown in Fig. 8C, sand each side of the lead until you get a chisel edge. Fig. 8D illustrates the chisel point. Chisel points give sharper lines than conical points, and draftsmen sometimes prefer chisel points for drawing straight lines.

A mechanical pointer, which saves much time and energy, can easily be obtained. It works better with loose leads than with wood pencils. Mechanical pointers assure you a perfect point at all times.

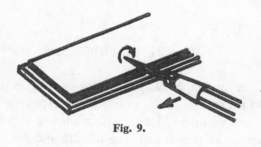

Fig. 9.

ERASERS AND ERASING SHIELDS

An eraser is used only to make changes in a drawing or to improve the design. A soft eraser, not an art-gum eraser, is usually used for erasing pencil marks or lines. A special ink eraser or steel eraser is usually used for erasing ink marks or lines. When a drawing requires a great deal of erasing, electric erasing machines are sometimes used.

Erasing a specific line on a crowded part of a drawing often causes trouble. If you use an erasing shield, illustrated in Fig. 10, you will erase only what you want to erase. The erasing shield must be held firmly when you are erasing.

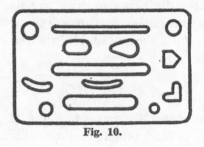

Fig. 10.

DUSTING BRUSH

Dusting brushes used by draftsmen have natural and synthetic bristles to help keep drawings clean and to eliminate erasure particles and other dust. Erasers in powder form may be sprinkled on a drawing and wiped off with a dusting brush. Erasing powder removes graphite dust with less effort than an eraser does.

SCALES

Every line must have a definite measurement. Ordinarily a 12-inch ruler is used for these measurements. However, many objects will not fit on the drawing paper and must be drawn in smaller size. The draftsman uses scales to draw **large objects** in smaller sizes and to draw **small objects** much larger than they actually are. He may also draw objects to their actual dimensions.

Many types of scales are available to the draftsman. Some scales will enable him to do the three kinds of measuring just described while others permit him to do only one or two kinds of measuring. All scales may be obtained in wood, in wood covered with plastic, or in all plastic.

The most widely used scale is the **architect's scale,** illustrated in Fig. 11. This scale enables

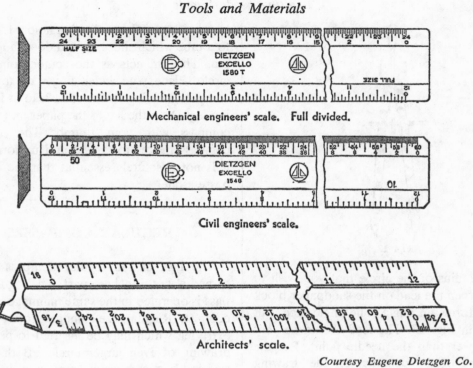

Mechanical engineers' scale. Full divided.

Civil engineers' scale.

Architects' scale.

Courtesy Eugene Dietzgen Co.

Fig. 11.

you to draw dimensions given in feet in lengths of inches. Some architect's scales contain several different scales. Architect's scales will help you to draw to the following scales: **3 inches for every foot; 1½ inches for every foot; 1 inch for every foot; ½ inch for every foot; and 3/16 of an inch for every foot.**

Fig. 11 also shows the **mechanical engineer's scale,** which is most helpful in drawing machinery and small machine parts. It has the ordinary 12-inch ruler length, with each inch divided into 32nds of an inch. The same scale may also have each 1-inch length marked to equal a ½-inch length. This is called a **half scale.** The scale may also have each 1-inch length marked to equal a ¼-inch length or a ⅛-inch length. These scales are known respectively as a ¼ **and ⅛ scale.**

The **civil engineer's scale,** also illustrated in Fig. 11, is used for map and chart work. This scale allows you to draw feet in lengths of inches as in the architect's scale, but each division on the scale is divided into ten parts. Map dimensions appear most frequently in decimals, as 350.25′, which means 350 feet and 25 hundredths of a foot. The .25 of a foot also equals

¼ of a foot or 3 inches. Therefore, the decimal actually is read as 350 feet and 3 inches. The civil engineer's scale will help you to mark off this dimension easily and accurately.

A scale should never be used to draw lines. Scales are only used for measuring. The markings on the edges of the scale prevent the drawing of a straight, sharp line.

COMPASSES

Almost all drawings include circles or parts of circles known as arcs. A compass is used to draw these circles and arcs. A small compass known as a **bow compass,** shown in Fig. 12, is used to draw small circles. Compasses have fittings which are inserted so that you can use them for drawing circles with pencils or ink. The compass has two legs. One leg holds a sharp-pointed piece of steel which serves as the center of the circle. The other leg contains the lead holder or ink container. The better compasses have a wheel between the legs, not outside the legs. The compass is operated by first turning the wheel between the legs. The distance between the legs, measured on the scale,

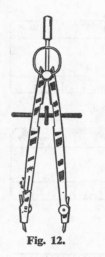

Fig. 12.

gives the radius of the circle that you wish to draw. Sharpen the lead on the sandpaper block so that it forms an angle. The angle faces out as shown in Fig. 13. The lead should be very slightly lower than the needle point. Put the sharp point of the compass on the drawing paper and turn the compass with the thumb and index finger.

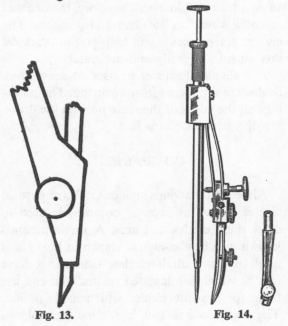

Fig. 13.   **Fig. 14.**

DROP-BOW COMPASSES

When a draftsman has to draw a great many small circles of the same size, he uses a **drop-bow compass.** Fig. 14 is a drawing of a drop-bow compass. The lead or ink holder is attached to a tube which slides up and down a long steel pin. The pin acts as the center point of the circle. Place your index finger on top of the compass, holding up the other lead or ink-holding leg. Drop the leg to the paper and twirl the compass with your thumb. The drop-bow compass makes for speed and uniformity but it is not absolutely essential that the beginner possess one.

REGULAR COMPASSES

The legs of the regular compass may be spread about 5 inches apart. The regular compass is operated in the same manner as the bow compass. Regular-size compasses often have an extra leg which may be inserted to permit the drawing of even larger circles. Both regular-size and bow compasses come in various qualities and the kind of compass that you buy will depend upon your needs and the amount of money that you wish to spend.

BEAM COMPASSES

A **beam compass,** shown in Fig. 15, is used to draw circles that have very large radii. The compass consists of two legs, with a round or square rod 10 to 12 inches long going through both legs. One leg is used for the center of the circle and the other for the lead. Extra rods may be attached if the draftsman wants to draw even larger circles. The legs move along the rod and are tightened by turning a screw against the rod from the top.

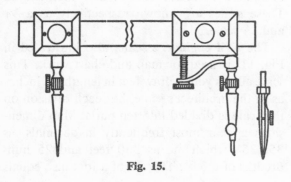

Fig. 15.

DIVIDERS

Dividers are used to mark off equal divisions on a line or to transfer a distance from one part of a drawing to another. Dividers have two legs, each containing a steel needle point. The ordinary divider used for general purposes is shown in Fig. 16A and the **bow divider,** used for working with small distances and containing a center screw, is shown in Fig. 16B.

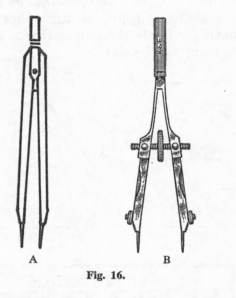

A   B

Fig. 16.

A divider is used the same way that a compass is used. When you work with dividers, you must be careful not to make large holes when inserting the divider points into the paper.

Fig. 17.

In addition to the two types of dividers which we have been discussing, draftsmen also use a third type known as **proportional dividers.** Proportional dividers, shown in Fig. 17, are used to establish the relationship between short lines and long lines, between radius and circumference, between feet and meters, as well as many other proportional relationships. They may also be used to make a drawing or part of a drawing an exact number of times larger. For example, a drawing done in a scale of 1 inch to a foot could be enlarged to a scale of 2 inches to a foot with proportional dividers.

RULING PENS

Many drawings are done in ink because ink does not rub off, because ink lines are distinct, and because ink drawings reproduce more clearly than pencil drawings. **Ruling pens** or special ink pens are used to draw lines in ink. The best ruling pen consists of a piece of stainless steel cut out down the center to form two blades. The points of the blades are rounded to form nibs. The piece of stainless steel is attached to a metal or wooden handle. The nibs should be rounded, as illustrated in Fig. 18. Tungsten carbide welded to the nibs helps them to keep their shape. The best pens come properly shaped right from the factory. The poorer ones may have to be rounded on an oilstone. Fig. 19 is a drawing of a ruling pen. Notice that a thumbscrew, threaded against one blade, rests against the inside of the other blade. The screws are turned to spread the blades apart or bring them closer together. Never bring the blades so close that they press against each other.

The ruling pen must be cleaned quite often because the India ink used with the pen dries rapidly. Ammonia or any commercial cleaner

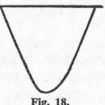

Fig. 18.

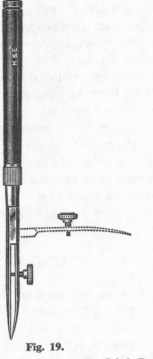

Fig. 19.

Courtesy Keuffel & Esser Co.

nibs with an ink holder. Fig. 20 shows a syringe-like holder attached to the bottle top. Another type of bottle stopper with a trough bent at the end is also available for filling. The ink should not be more than ¼ of an inch high. India ink dries quickly on exposure and therefore it is a good idea to keep the bottle closed when it is not being used. If the ink in the pen dries, dip the pen in the bottle and then gently wipe off the sides of the pen. Fig. 21 shows the proper way to fill the pen. The pen should be held at an angle to the paper, but flat against the T square or triangle. Do not press the pen against the triangle or T square.

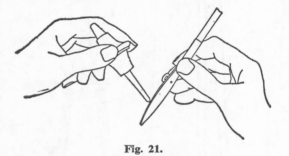

Fig. 21.

may be used to clean a pen. Variations in the nib attachment have been introduced in order to simplify the job of cleaning the pen. These variations, shown in Fig. 19, can affect the relationship between the nibs. These improvements are contained in only the finest and most precisely made pens.

Compasses, bow compasses, dividers, and ruling pens, together with extra screws, legs, and leads, are sold in sets. Buy only what you need, but buy good-quality materials.

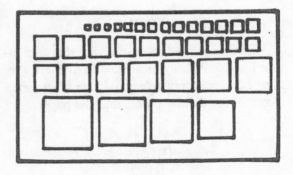

Fig. 20.

INKS

When you are required to make a drawing with ink, you should use a black waterproof ink that is frequently known as India ink. The ruling pen is filled by putting some ink between the

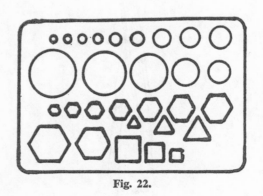

Fig. 22.

TEMPLATES

In the course of his work the draftsman is required to use many geometric shapes, including circles, squares, ellipses, and rectangles. Plastic sheets with these various shapes accurately cut out may be purchased. Many industries use standard symbols for various objects. **Templates,** shown in Fig. 22, with these symbols can also be purchased. The draftsman finds these templates to be very helpful in his work. Used properly they can save you time and energy.

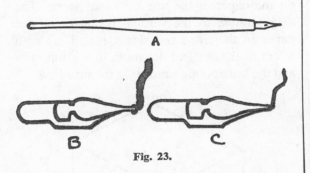

Fig. 23.

LETTERING PENS

Pencils are, of course, used for lettering in pencil drawings. However, lettering on ink drawings requires the use of a special **lettering pen.** Fig. 23A illustrates a commonly used lettering pen point and holder. Pen points for making wide or narrow strokes may be obtained. Fig. 23B illustrates the type of point that should be used to make broad, wide strokes. Fig. 23C illustrates a fine steel pen point which can be used to make thin, delicate-stroked letters.

Fig. 24.

Fig. 24 shows a mechanical type of lettering pen. The same thickness is obtained regardless of the direction of the stroke of the letter. This particular mechanical pen, one of the latest types, is called a **Rapidograph** and can be used for drawing as well as lettering. Each of the five different-size pens available, numbers *00, 0, 1, 2,* and *3,* draws a different line thickness. The Rapidograph contains a large amount of India ink so that it does not have to be filled as often as the ordinary ruling pen. The Rapidograph may also be used with lettering templates made by the same manufacturer. Many of the illustrations for this book were made with Rapidograph pens.

Leroy and Wrico lettering devices remove most of the hand-lettering problems. A variety of types and sizes are available, as well as templates for use with the lettering set. The skillful use of these devices requires practice. Hand lettering, however, still has a quality not apparent in more mechanically formed letters.

CHAPTER TWO

ORTHOGRAPHIC PROJECTION

Artists draw objects for others to look at and enjoy. Draftsmen, however, are engaged in drawing objects to be constructed by others. Although several drawing methods have been devised for this purpose, our study will concern itself with orthographic projection, the method which is most commonly used.

If you want a square piece of wood cut, simply mark off the size of the sides, which happens to be 12 inches in this case. See Fig. 25A. If

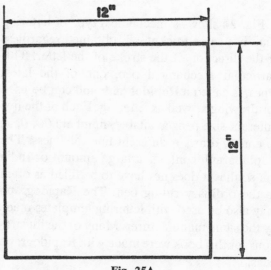

Fig. 25A.

you wanted a rectangular piece of wood to be cut to certain specific dimensions, such as 6 inches long, 4 inches wide, and 8 inches deep, you would simply give the dimensions to a carpenter. But how would you draw a complicated machine, a house, a sofa, or a bookcase for construction. The orthographic-projection method shows you how to draw an object which has depth on a sheet of paper which has only length and width. Naturally, it will be necessary to discuss the idea behind the orthographic-projection method before any drawing is attempted.

Ordinarily the words *length*, *width*, and *height* are used to describe the basic dimensions of an object. However, in drafting, the definitions of these words are slightly different. In this book, **width, height,** and **depth** are used to show the basic dimensions. Figs. 25B and 25C show their relative positions. As Figs. 25B and 25C indicate, width refers to the horizontal distances on the paper in the top and front views. The height refers to the vertical distances on the paper in the front and side views. The depth refers to the vertical distances in the top view and the horizontal distances in the side view.

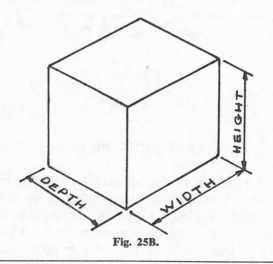

Fig. 25B.

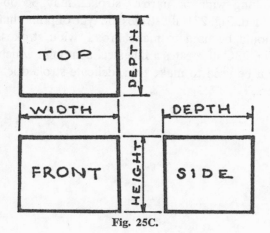

Fig. 25C.

FRONT VIEW

Study the block used in Fig. 26. Imagine a sheet of glass standing up straight between your eyes and the block. The sheet of glass represents the sheet of drawing paper. It has only width and height. The block, however, has depth as well as width and height. When you look straight at the front of the block you do not see the top or the sides. You see only the outline, *A, B, C, D*. Imagine lines going straight from these points to the sheet of glass. Connect these points on the glass. You now have the outline of the front of the block, or the front view. Incidentally, you follow the path of a line by going from one adjacent point to another. For example, the front view would be read as *A, B, C, D* and not *A, B, D, C*.

TOP VIEW

Now imagine a sheet of glass between your eyes and the block as you look straight down on the block. See Fig. 27. All you see is the outline, *A, D, F, E*. You do not see the front or the sides. Imagine lines going from the points *A, D, F, E* to the sheet of glass. Connect the points on the glass. You now have the outline of the top or the top view.

SIDE VIEW

In order to have a complete picture of the block, you must know what the sides look like. Again, look straight at the side, as shown in Fig. 28, with the sheet of glass between your eyes and the block. You see only points *D, C, G, F*. Imagine straight lines projecting from these points to the glass. Connect the points and you have the outline of the side or the side view.

PLACING OF THE VIEWS

The top, front, and side views give us a complete picture of the block. Fig. 29 illustrates the proper placement of the views on paper. The front view comes directly under the top view. The side view stands along the right side of the front view. This arrangement helps us to see the relationship between the views. The top view and the front view both contain the width of the block, *A to D*. The side view and the top view both contain the depth of the block, *D to F*. The front view and side view both contain the height of the object. It will be simpler to draw the views with a knowledge of these relationships.

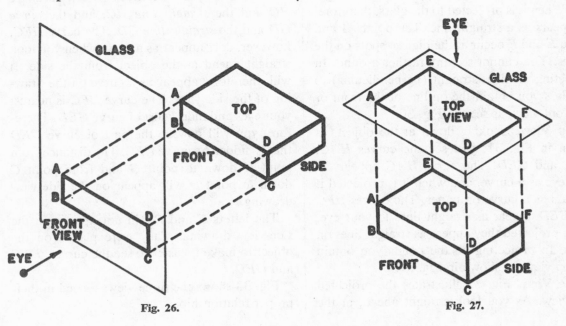

Fig. 26.　　　　　　　　　　　　　　Fig. 27.

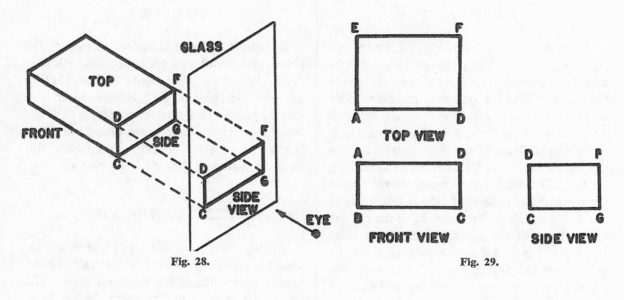

Fig. 28.   Fig. 29.

ROUND SURFACES AND CURVED LINES

Front View. Our discussion up to this point has been based on blocks consisting of flat surfaces and straight lines. Suppose you have an object consisting of rounded surfaces and curved lines, as illustrated in Fig. 30. In the front view you see points *A, B, D, E, F, C*. The line *D to B* is actually curved. When you look straight at it, however, it appears as a straight line. When it is projected to the glass, therefore, it appears as a straight line. The points *A* and *B* and *E* and *F* occur at the highest spots on the curves. You cannot see any farther around. In projecting the lines straight (perpendicular) to the glass you obtain the outline of the front or the front view, as seen in Fig. 30.

Top View. Looking down at the object, as shown in Fig. 31, you see the curves *HAC, HFC,* and *GED.* The curve *HAC* appears to your eye as a curve, and when it is projected it appears as a curve on paper. The curves *HFC* and *GED* appear as straight lines to your eye. When projected they appear as straight lines on paper. In projecting all the points you obtain the top view, as shown in Fig. 31.

Side View. Fig. 32 illustrates the projected side view. As you look straight ahead, at the side, the curve *HAC* is nearest your eyes and hides the straight line *HC.* In the same manner curve *GBD* is nearest your eyes and hides line *GD.* But the curves *HAC* and *GBD* appear as straight lines when you look straight ahead. So, in drawing the side view, the curves *HAC* and *GBD* will become straight lines. Actually the points *A* and *B* are omitted in the side view. They were placed on the object merely to make it easier to see the difference between the *curve HC* and the *straight line HC,* and the *curve GD* and the *straight line GD*. The curve *HFC,* however, does appear as a curve when you look straight ahead at the object from the side. It will, therefore, appear as a curve on the drawing of the side view. The curve *HFC* is nearest your eyes and hides most of curve *GED.* Therefore, you will not see the part of curve *GED* that is hidden by curve *HFC.* The distance from point *H* down to point *G* and from point *C* down to point *D* will appear on the side-view drawing.

The letters *E* and *F* will not appear on the side-view drawing. They were placed on the object to make it easier to see the curves, *HFC* and *GED.*

Fig. 33 shows the three views placed in their proper relationship.

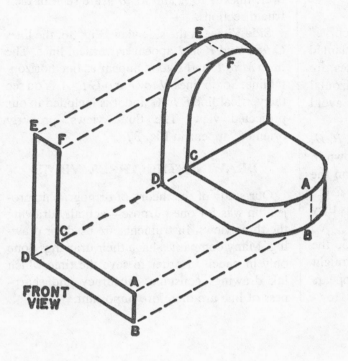

Fig. 30.

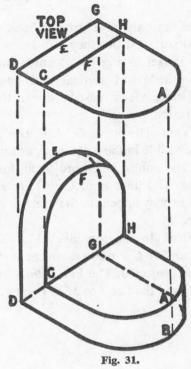

Fig. 31.

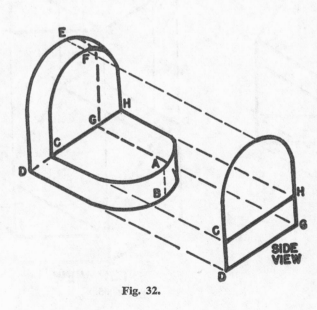

Fig. 32.

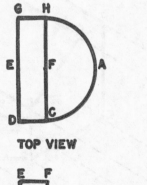

TOP VIEW

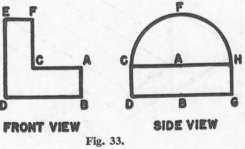

FRONT VIEW   SIDE VIEW

Fig. 33.

SLANTED LINES

Objects not only have straight lines, curved lines, and surfaces, but they also have slanted lines, as shown in Fig. 34. Slanted lines are often confused with vertical and horizontal lines. All three views must be drawn to avoid such confusion.

Front View. The points *G, A, E, D, C, B, H,* and *F* are visible in the front view, as shown in Fig. 34. The points, when projected, form the front view. The lines *A to G* and *H to F* are projected as they appear to the eye, not as they really are.

Top View. In the top view, Fig. 35, the slanted lines *B to C* and *J to K* appear as straight lines when projected. The line *B to H* appears as point *B* and the line *J to M* as point *J*. How-ever, lines *H to F* and *M to F* are seen in their true directions.

Side View. In the side view, Fig. 36, the lines *C to B* and *K to J* appear as vertical lines. The lines *M to F* and *F to H* appear as one horizontal line. So do lines *N to G* and *G to A*. You see the vertical line *F to G* and it is included in our projected view. The three views, properly oriented, appear in Fig. 37.

DRAWING THE THREE VIEWS

Our study of the theory of orthographic projection was for one purpose—actually drawing the three views. Instruments are used for drawing. Many companies have their drawings done only in pencil in order to save the time which ink drawings consume. Accuracy and sharpness of line are therefore important.

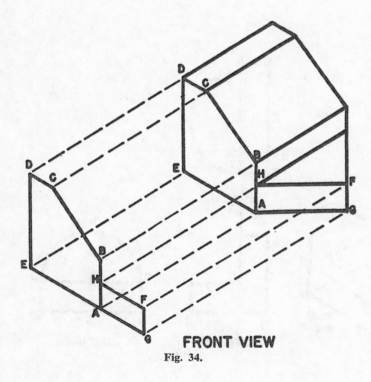

FRONT VIEW

Fig. 34.

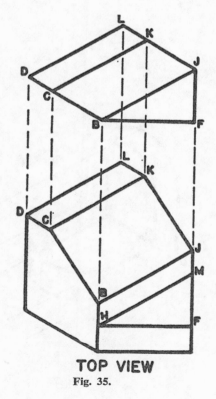

TOP VIEW

Fig. 35.

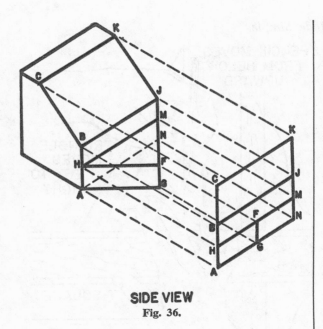

SIDE VIEW
Fig. 36.

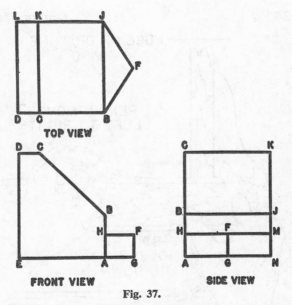

TOP VIEW

FRONT VIEW

SIDE VIEW

Fig. 37.

SIZE OF PAPER

It is wise to use the standard-size sheets used in industry. The American Standards Association recommends the following sizes:

Table I

LETTER DESIGNATION	WIDTH	LENGTH
A	8½	11
B	11	17
C	17	22
D	22	34
E	34	44

Table II

LETTER DESIGNATION	WIDTH	LENGTH
A	9	12
B	12	18
C	18	24
D	24	36
E	36	48

These sizes are based on multiples of 8½ × 11 inches in Table I and 9×12 inches in Table II. American industry uses both sizes and therefore both are considered standard.

The beginner should start with a heavy vellum, either 8½″×11″ or 11″×17″.

Begin by fastening the paper to the drawing board with a small piece of masking tape on each corner of the paper, as shown in Fig. 2,

Chapter One. The sheet should be closer to the left side of the board because the T square slides up and down the left side of the drawing board. Place the T square on the lower part of the board and try to fasten the paper down so that the bottom edge of the paper lines up with the T square. The paper will then rest squarely on the board.

All the lines should be drawn lightly with a 4H pencil. Draw the margins first. Using 11″×17″ paper, measure down ½ inch from the top edge of the paper. Place the T square along the left side of the board and slide it up to the ½-inch mark. Press down on the T square to keep it from moving. Make sure it hugs the left side of the board. Hold the pencil so that it slants up to the right. Make sure the pencil point is at the place where the T square and paper meet, as shown in Fig. 38. Draw the margin line across the top of the paper. From the bottom edge of the paper, measure up ½ inch and draw the bottom line.

Measure in ½ inch from the right edge and ½ inch from the left edge of the paper. Use the triangle for drawing vertical lines. Place it on the drawing paper so that it rests along the top edge of the T square, as shown in Fig. 39. Move the triangle along until it meets the ½-inch mark and connect the two horizontal margin lines. Do not draw the line to the point of the triangle. If the triangle is too short, move the T square and triangle up and continue the line.

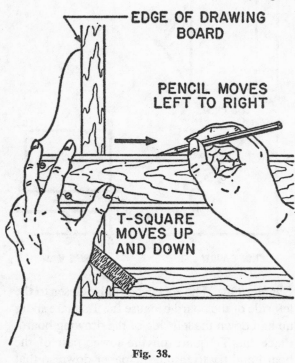

EDGE OF DRAWING BOARD

PENCIL MOVES LEFT TO RIGHT

T-SQUARE MOVES UP AND DOWN

Fig. 38.

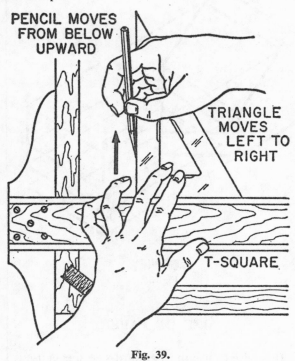

PENCIL MOVES FROM BELOW UPWARD

TRIANGLE MOVES LEFT TO RIGHT

T-SQUARE

Fig. 39.

TITLE BLOCK

Draw two lines 4½"×2" in the lower right-hand corner to form a box with the two margin lines. This box is called a **title block** and is shown in Fig. 40. The title block contains the title of the drawing, the draftsman's name, the date, and the scale of the drawing. In commercial practice the title block may already be printed on the drawing sheet and may require additional information.

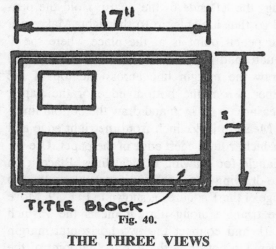

17"

TITLE BLOCK

Fig. 40.

THE THREE VIEWS

Let us try to draw the three views of the object shown in Fig. 41. We have a three-di-

mensional sketch, with given dimensions, to use as a basis for our drawing.

Remember to draw all three views together. Do not finish one view before going to another.

A study of the sketch will help in determining where to place the three views. Do not crowd them. Space must be allowed for dimensions and notes. The finished drawing should also be pleasant to look at. The drawing sheet measures 11"×17". A ½-inch margin leaves 10"×16" of working space. The title block is placed two inches from the bottom margin. The right side view must not be too close to the title block. All three views could, therefore, be above the title block, or to the left of the title block. This is shown in Fig. 40. The area above the title block could then contain general notes.

There should be space around each view for dimensions and notes. Our views will be drawn in full scale. In full scale, the top view will take up 2 inches in height and the front view will take up 3 inches in height. Since our margins take up ½ of an inch on top and ½ of an inch on the bottom, and our sheet is 11 inches long, we still have 10 inches in which to place these two views. Since the two views take up 5 inches, we still have 5 inches of space with which to

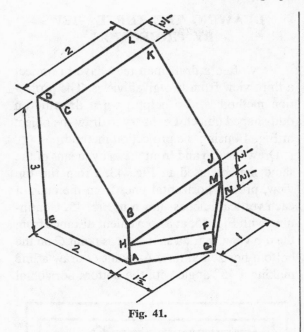

Fig. 41.

work. By putting the two views 2 inches apart, we have 1½ inches left on the top and 1½ inches left on the bottom.

Looking from left to right, we see that 11½ inches remain for our three views since we have already allowed a margin of ½ of an inch on each side and 4½ inches for the title block. This is also true because all three views will be drawn to the left of the title block. The front and side views will take a total of 5½ inches, which leaves 6 inches. The left edge of the front view could be placed 2 inches from the left-hand margin and there could be 2 inches between the front and side views. That would leave 1½ inches to the title block.

As you can see from the discussion above, the draftsman must do a great deal of planning before he can begin to make a drawing. In order to obtain the outline blocks for the three views, shown in Fig. 42, start by measuring down 1¼ inches from the top margin, then 2 inches from that spot, 2 more inches from that spot, and finally 3 inches more. At these four points, draw light horizontal lines with the T square. The bottom two lines should go to within 1 inch of the title block. The two lines will be for the front and side views. The two top lines should be shorter.

From the left-hand margin, measure consecutively 2 inches, 3½ inches, 2 inches, and finally 2 inches. With the triangle resting on the T square, draw vertical lines at these points. The first two vertical lines on the left should extend from beyond the top horizontal line to beyond the bottom horizontal line. The second two vertical lines should extend beyond the bottom two horizontal lines only. You now have the outline blocks for the three views, as shown in Fig. 42.

Working with Fig. 43, where the top horizontal line crosses the first vertical line on the left, which is point *L*, measure ½ inch to the right. That gives us point *K*. Now measure 2 inches horizontally from point *L*, which gives us point *J*. Draw vertical lines to cross the front view from points *K* and *J* parallel to the first vertical line from point *L*.

The points *D*, *C*, and *B* can now be found where the second horizontal line crosses the

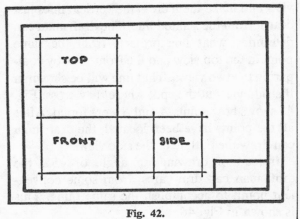

Fig. 42.

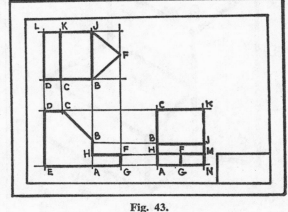

Fig. 43.

vertical lines just drawn from points *L*, *K*, and *J*. From the end of the horizontal line beyond point *J*, in the top view, measure down 1 inch, which locates point *F*. Connect point *F* with points *B* and *J*. The lines defining the top view may now be darkened.

In the front view, Fig. 43, points *D* and *C* occur where the vertical lines for the top view cross the uppermost horizontal line of the front view. Point *G* in the front view occurs at the crossing of the bottom horizontal line of the front view and the vertical line from point *F* in the top view. Point *F* in the front view can be found by measuring up ½ inch from point *G*. Point *B* in the front view can be found by first measuring up ½ inch from point *F* and then drawing a horizontal line. Where this horizontal line crosses the vertical line from point *B* in the top view, you will have point *B* in the front view. Point *H* in the front view will be found ½ inch below *B* in the front view.

The various heights of the object are the same in the front and side views. The heights have already been determined for the front view. Carry them across to the side view by drawing horizontal lines from points *B* and *F*. Points *D* and *E* were previously drawn when the three views were first laid out. Point *F* in the side view is 1 inch from point *M*.

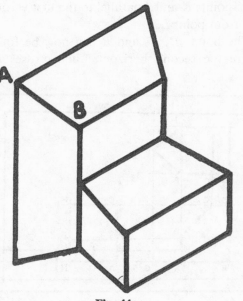

Fig. 44.

DRAWING THE THREE VIEWS BY PROJECTION

Very often a draftsman may have to project a third view from two given views. The projection method is also helpful when drawing an odd-shaped object. Let us try to draw the object in Fig. 44 using the projection method.

Draw the top and front views by means of the same method used in Fig. 43. From the top view, project horizontal lines from the ends of each vertical line, as shown in Fig. 45. Continuing with Fig. 45, at a convenient distance from the top view, mark a spot such as point *O* on the bottom horizontal line. At point *O* draw a line making a 45° angle with the bottom horizontal

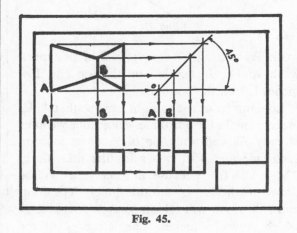

Fig. 45.

line. This slanting line will cross the other horizontal lines projecting from the top view. Make dots where the lines cross. From these dots, draw vertical lines down.

Now project horizontal lines from the front view and make them cross the vertical lines just drawn from the slanted line. The draftsman can determine what line projects from the same point in the top view and the front view by comparing the two views. That line will be shown in the side view at the spot where they cross. Fig. 45 shows how points *A* and *B* were found. After all the points have been located, the draftsman can draw the outline of the side view.

Instead of drawing an angle of 45°, the draftsman can draw arcs, from some convenient point, to accomplish the same thing. This is shown in Fig. 46.

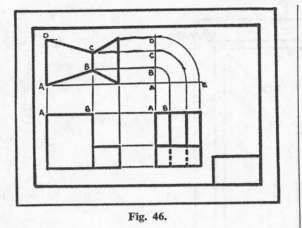

Fig. 46.

Working with Fig. 46, after drawing horizontal projection lines from the top view, drop a vertical line perpendicular to these horizontal lines. With point *A* as a center and with points *B, C,* and *D* as radii from *A*, draw arcs to the line *AE*. From line *AE* drop vertical lines. Where these lines cross the horizontal lines from the front view you obtain the points for drawing the shape of the side view.

PROJECTING A SLANTED SURFACE WITH A HOLE IN IT

If an object has a slanted surface with a hole in it, the use of the projection method is extremely helpful. In drawing the three views of the object shown in Fig. 47, start by drawing the top and front views, as shown in Fig. 48.

Now project lines horizontally from the top view, draw the 45° angle, and project lines down from the angle line. Then project lines from the front view and cross the vertical lines. Draw the outline of the object. The hole in the side view still remains to be drawn. In the top view it appears as a circle. In the side view it will appear in an oval shape, called an ellipse.

In order to draw the ellipse various points in the oval shape, as shown in Fig. 48, must be found. The points are then connected to form the curved shape of the ellipse. To find these points, pick any place on the circle in the top view and draw a projecting line to the 45° angle, then drop down to the side view. From the same place on the circle in the top view, project a line down to the slanted line in the front view. Where this projected line crosses the slanted line in the front view, draw a horizontal projection line, making it cross, in the side view, the vertical line that came from the same place in the circle in the top view. The crossing of the two lines gives us one point of the curve of the ellipse. By repeating this procedure from different spots on the circle in the top view, many points of the elliptical curve can be found. The points can be connected by drawing a curve with the French curve. For the completed ellipse, see Fig. 49.

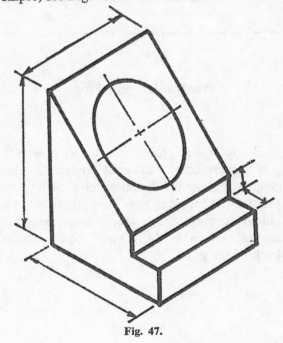

Fig. 47.

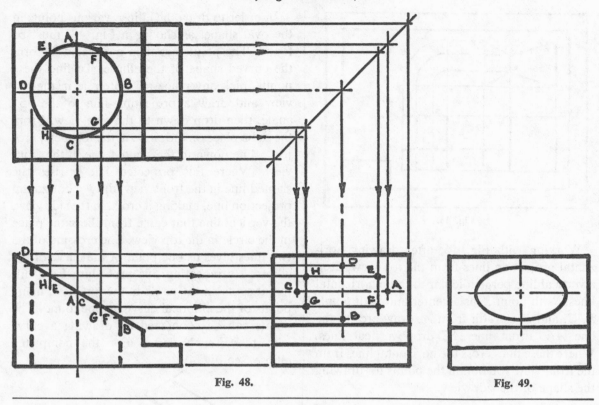

Fig. 48.   Fig. 49.

Practice Exercise No. 1

The following problems will test how well you understand the material covered in this chapter. Follow the directions given with each problem. After you have completed the required drawings check your solution with the answers shown in the answer section at the back of the book.

PROBLEM 1: Draw right side view.

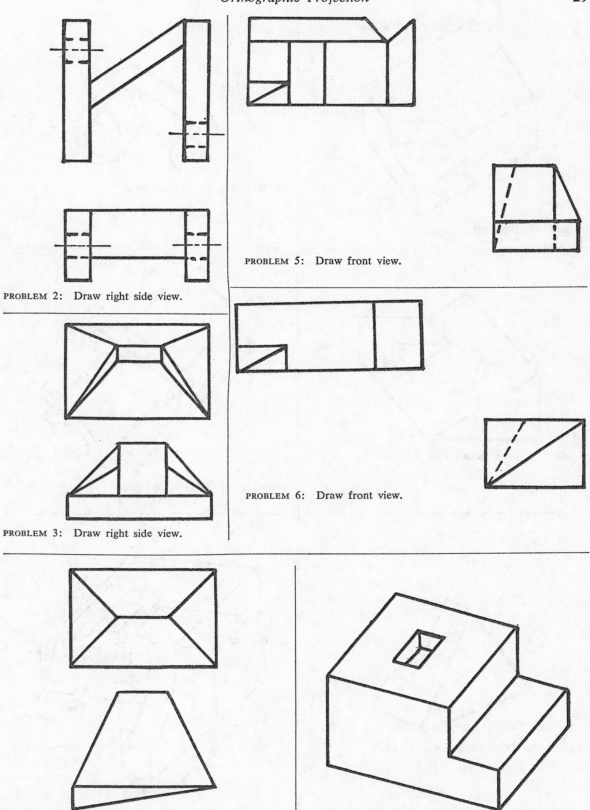

PROBLEM 2: Draw right side view.

PROBLEM 3: Draw right side view.

PROBLEM 4: Draw left side view.

PROBLEM 5: Draw front view.

PROBLEM 6: Draw front view.

PROBLEM 7: Draw three views.

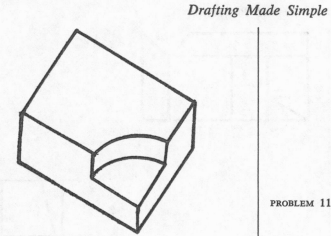

PROBLEM 8:   Draw three views.

PROBLEM 11:   Draw three views.

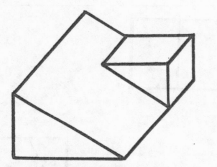

PROBLEM 9:   Draw three views.

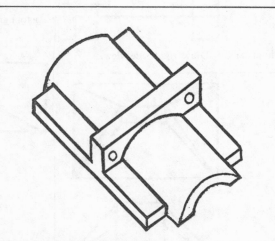

PROBLEM 12:   Draw three views.

PROBLEM 10:   Draw three views.

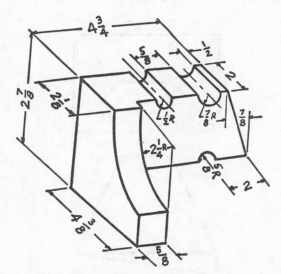

PROBLEM 13:   Draw three views to scale.

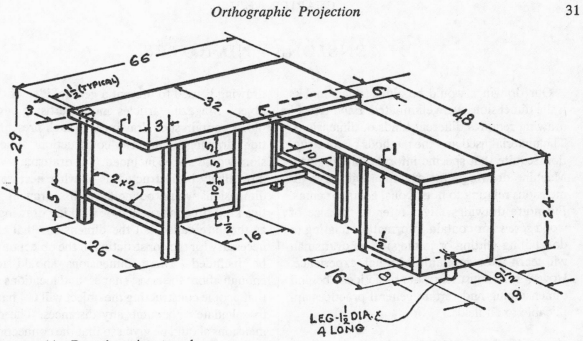

PROBLEM 14: Draw three views to scale.

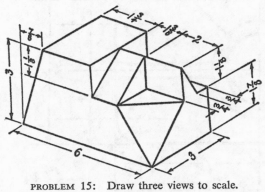

PROBLEM 15: Draw three views to scale.

CHAPTER THREE

DIMENSIONS AND NOTES

Our drawings would be almost meaningless if the dimensions were eliminated. Each type of drawing requires different kinds of dimensions. The material used and the methods of construction require that specific information be given. Machine drawings, for example, will have dimensions relating to nuts, bolts, gears, or cams. Furniture drawings might refer to the size of wood screws or contain information relating to dovetailing (fitting or joining). The draftsman will learn these specialties through experience. However, he must learn the basic ideas behind dimensioning and certain general practices applicable to all fields.

THEORY OF DIMENSIONING

The type and variety of dimensions depend upon the purpose of the drawing. Sometimes a drawing is used to present a general idea of the object. Magazine articles and advertising pamphlets contain such drawings. In such presentation drawings complete construction dimensions would not be included. The draftsman will most likely be instructed as to what necessary dimensions would be included in a drawing of this kind. In construction or working drawings, as they are called, all the dimensions that are necessary for the construction of the object must be included. These dimensions should tell enough about the sizes, shapes, and locations so that people constructing the object will not have to calculate or assume any distances. The dimensions should be given so that the connection between related dimensions can be readily seen.

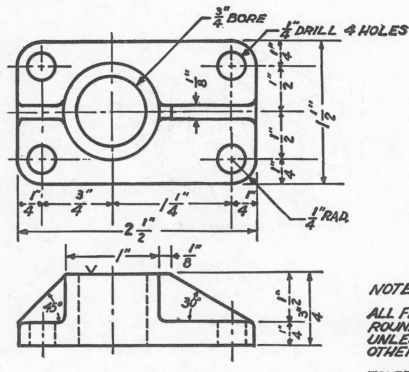

Fig. 50.

SIZE AND LOCATION DIMENSIONS

The dimensions must show the size of an object and the size of its various parts. As Fig. 50 indicates, size dimensions include length, width, and depth of objects, the radius or diameter of circles and curves, and the sizes of angles. Fig. 51 shows some examples of size dimensions.

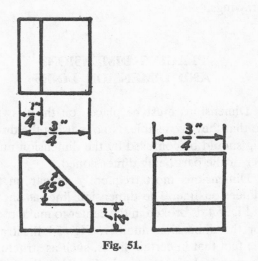

Fig. 51.

The draftsman uses **location** dimensions as well as **size** dimensions. Location dimensions show the position of several objects in one drawing, or the location of different parts of one object. Machines in a factory, buildings of an industrial plant, furniture in a room, water pipes in a house, or holes in a sieve require location dimensions when drawn for construction. Fig. 52 shows some examples of location dimensions.

TECHNIQUES IN DRAWING DIMENSIONS, EXTENSIONS, AND DIMENSION LINES

Dimensions are denoted by drawing extension lines at each end of the dimensioned part. These extension lines are drawn perpendicular to the outline of the object. A line with an arrow at each end, called the dimension line, meets the extension line. See. Fig. 50.

The extension lines do not touch the outline of the object. They extend a little beyond the dimension line. The extension and dimension lines should be drawn thinner than the lines of the object. See Fig. 50.

ARROWS

The arrows on the dimension line must be drawn so that the length of the arrow appears to be about three times its own width. This is shown in Fig. 53.

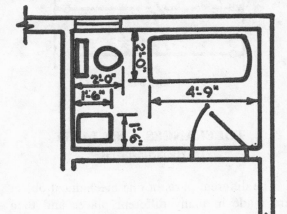

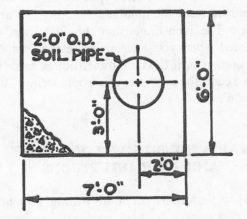

Fig. 52.

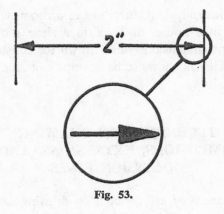

Fig. 53.

WRITING DIMENSION NUMBERS

Since the metric system is used in most foreign countries, dimensions for most foreign drawings are in decimal form. In the United States decimals are also used, but most of the time feet and inches are used. The mark (″) stands for **whole inches.** Three inches is written 3″. The mark also stands for parts of an inch. Three-quarters of an inch is written ¾″. If you use decimals in your drawings, you would write five-thousandths of an inch, for example, as .005″. Decimal dimensions are often found in machine drawings, where great accuracy in construction is required.

The mark (′) stands for **feet.** We write five feet as 5′-0″ and five feet three and three-quarter inches as 5′-3¾″. The hyphen should always appear between the numbers denoting feet and inches. The inch dimension should always be included when you have to write the foot dimension, even if the inch dimension is zero. In such a case, for example, you would write 6′-0″, never 6′.

VARIATIONS IN WRITING FOOT AND INCH DIMENSIONS

Many variations occur in writing foot and inch dimensions. In machine drawings, for example, the inch mark is not used if all the dimensions on the drawing are given in inches. The number appears without the mark, as 3¼, for 3¼ inches. Some industries specify that all

dimensions up to 72 inches must be written in inches. Dimensions 72 inches and larger are written in feet and inches. Drawings showing steel construction omit the inch mark in the foot and inch dimensions, as for example, 6′-0.

The draftsman will soon learn the general practices of the particular field in which he works. In many cases individual companies establish drafting practices for use on their own drawings.

PLACING DIMENSIONS AND DIMENSION LINES

Dimensions must be placed on the drawing so that they are legible. No one using the drawing should be confused by the dimension number or the part being dimensioned.

Dimensions most frequently appear on the dimension line. The dimension line, as shown in Fig. 54, is broken in the center to make room for the dimension number. Fig. 54 illustrates the fact that in certain fields, such as structural or chemical-plant construction, the dimension numbers appear over the dimension lines. Again, the draftsman will soon learn to conform with the general practices of his field.

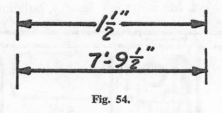

Fig. 54.

TOLERANCES AND LIMIT DIMENSIONING

The different parts of one mechanical object are made in many different places and then brought together and assembled. It is therefore important that parts should be interchangeable. In order to make that possible the person who makes each part must be allowed certain variations in the dimensions of the parts. Furthermore, the mechanic cannot make a part to the exact size, especially if the dimensions are extremely close, such as 6.326 inches. For these two reasons the machinist is allowed a certain leeway, which is called a tolerance.

When two nonthreaded parts, such as a shaft in a hole, fit into each other, both the size of the shaft and the size of the hole are limited. Let us say that the hole will be 1½″ in diameter. It cannot be less than 1½″ but it can be more. If it is more, the shaft will still fit into the hole. In this instance, we shall allow the machinist a leeway of .001″. That is, the hole may not be less than 1.500 inches, and it may not be more than 1.501 inches. The limit dimensions of the hole are, therefore, 1.500 and 1.501. The tolerance is .001.

The same procedure applies to the shaft. It must be less than 1.500 inches to fit into the hole. Let us assume that the shaft cannot be larger than 1.498 inches. We allow the machinist a .001-inch leeway, which means the shaft cannot be less than 1.497 inches. See Fig. 55.

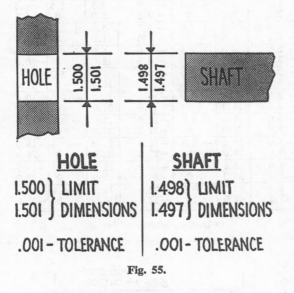

HOLE

1.500 } LIMIT
1.501 } DIMENSIONS

.001 - TOLERANCE

SHAFT

1.498 } LIMIT
1.497 } DIMENSIONS

.001 - TOLERANCE

Fig. 55.

Although a great variety of parts are mated or fitted to other mechanical parts, it has been possible to organize the closeness of all these fits into five basic types. For such information the draftsman should consult the American Standard B4.1–1955, *Preferred Limits and Fits for Cylindrical Parts,* published by the American Society of Mechanical Engineers and approved by the American Standards Association. Each type of fit has several classes, as follows:

1. **RC**–Running or sliding fit
   a. **RC 1-Close sliding fit**
   b. **RC 2-Sliding fit**—more clearance than RC 1
   c. **RC 3-Precision running fit**—for precession work at slow speeds.
   d. **RC 4-Close running fit**
   e. **RC 5** and **RC 6-Medium running fit**—for higher running speeds.
   f. **RC 7-Free running fit**—for minimum accuracy.
   g. **RC 8** and **RC 9-Loose running fits**—for tubes and cold-rolled shafts.
2. **LC**–Location Fits
   a. Locational **Clearance Fits**—for stationary parts, such as a spigot.
   b. **LT**-Locational **Transition Fits**—a compromise between **clearance** and **interference fits.**
   c. **LN**-Locational **Interference Fits**—for accuracy of location.
3. Force or Shrink Fits
   a. **FN1-Light drive fits**—for light assembly pressures.
   b. **FN2-Medium drive fits**—for ordinary steel parts.
   c. **FN3-Heavy drive fits**—for heavier steel parts.
   d. **FN4** and **FN5-Force fits**—for highly stressed parts.

The limit dimensions for each of these fits are in various tables. Fig. 56 is a table showing some of the running-fit dimensions. The column on the left gives the hole diameters. To determine the tolerances for a running fit of a hole whose diameter is 5.250 inches, you look down the left column and find it between 4.73 and 7.09. You decide to use an *RC2*-class fit. Looking to the right under the *RC2*-class column, you find that the standard limits for the hole are from 0.0000 to 0.0010 inches, and for the shaft from −0.0006 to −0.0013. The limits for the hole are added to the basic hole size, as follows:

5.2500+0.0000=5.250 (smallest hole)
5.2500+0.0010=5.251 (largest hole)

RUNNING AND SLIDING FITS

Nominal Size Range Inches Over — To	Class RC 1 Limits of Clearance	Class RC 1 Standard Limits * Hole H5	Class RC 1 Standard Limits * Shaft g4	Class RC 2 Limits of Clearance	Class RC 2 Standard Limits * Hole H6	Class RC 2 Standard Limits * Shaft g5	Class RC 3 Limits of Clearance	Class RC 3 Standard Limits * Hole H6	Class RC 3 Standard Limits * Shaft f6	Class RC 4 Limits of Clearance	Class RC 4 Standard Limits * Hole H7	Class RC 4 Standard Limits * Shaft f7
0.04– 0.12	0.1 / 0.45	+0.2 / 0	–0.1 / –0.25	0.1 / 0.55	+0.25 / 0	–0.1 / –0.3	0.3 / 0.8	+0.25 / 0	–0.3 / –0.55	0.3 / 1.1	+0.4 / 0	–0.3 / –0.7
0.12– 0.24	0.15 / 0.5	+0.2 / 0	–0.15 / –0.3	0.15 / 0.65	+0.3 / 0	–0.15 / –0.35	0.4 / 1.0	+0.3 / 0	–0.4 / –0.7	0.4 / 1.4	+0.5 / 0	–0.4 / –0.9
0.24– 0.40	0.2 / 0.6	+0.25 / 0	–0.2 / –0.35	0.2 / 0.85	+0.4 / 0	–0.2 / –0.45	0.5 / 1.3	+0.4 / 0	–0.5 / –0.9	0.5 / 1.7	+0.6 / 0	–0.5 / –1.1
0.40– 0.71	0.25 / 0.75	+0.3 / 0	–0.25 / –0.45	0.25 / 0.95	+0.4 / 0	–0.25 / –0.55	0.6 / 1.4	+0.4 / 0	–0.6 / –1.0	0.6 / 2.0	+0.7 / 0	–0.6 / –1.3
0.71– 1.19	0.3 / 0.95	+0.4 / 0	–0.3 / –0.55	0.3 / 1.2	+0.5 / 0	–0.3 / –0.7	0.8 / 1.8	+0.5 / 0	–0.8 / –1.3	0.8 / 2.4	+0.8 / 0	–0.8 / –1.6
1.19– 1.97	0.4 / 1.1	+0.4 / 0	–0.4 / –0.7	0.4 / 1.4	+0.6 / 0	–0.4 / –0.8	1.0 / 2.2	+0.6 / 0	–1.0 / –1.6	1.0 / 3.0	+1.0 / 0	–1.0 / –2.0
1.97– 3.15	0.4 / 1.2	+0.5 / 0	–0.4 / –0.7	0.4 / 1.6	+0.7 / 0	–0.4 / –0.9	1.2 / 2.6	+0.7 / 0	–1.2 / –1.9	1.2 / 3.6	+1.2 / 0	–1.2 / –2.4
3.15– 4.73	0.5 / 1.5	+0.6 / 0	–0.5 / –0.9	0.5 / 2.0	+0.9 / 0	–0.5 / –1.1	1.4 / 3.2	+0.9 / 0	–1.4 / –2.3	1.4 / 4.2	+1.4 / 0	–1.4 / –2.8
4.73– 7.09	0.6 / 1.8	+0.7 / 0	–0.6 / –1.1	0.6 / 2.3	+1.0 / 0	–0.6 / –1.3	1.6 / 3.6	+1.0 / 0	–1.6 / –2.6	1.6 / 4.8	+1.6 / 0	–1.6 / –3.2
7.09– 9.85	0.6 / 2.0	+0.8 / 0	–0.6 / –1.2	0.6 / 2.6	+1.2 / 0	–0.6 / –1.4	2.0 / 4.4	+1.2 / 0	–2.0 / –3.2	2.0 / 5.6	+1.8 / 0	–2.0 / –3.8
9.85– 12.41	0.8 / 2.3	+0.9 / 0	–0.8 / –1.4	0.8 / 2.9	+1.2 / 0	–0.8 / –1.7	2.5 / 4.9	+1.2 / 0	–2.5 / –3.7	2.5 / 6.5	+2.0 / 0	–2.5 / –4.5
12.41– 15.75	1.0 / 2.7	+1.0 / 0	–1.0 / –1.7	1.0 / 3.4	+1.4 / 0	–1.0 / –2.0	3.0 / 5.8	+1.4 / 0	–3.0 / –4.4	3.0 / 7.4	+2.2 / 0	–3.0 / –5.2
15.75– 19.69	1.2 / 3.0	+1.0 / 0	–1.2 / –2.0	1.2 / 3.8	+1.6 / 0	–1.2 / –2.2	4.0 / 7.2	+1.6 / 0	–4.0 / –5.6	4.0 / 9.0	+2.5 / 0	–4.0 / –6.5
19.69– 30.09	1.6 / 3.7	+1.2 / 0	–1.6 / –2.5	1.6 / 4.8	+2.0 / 0	–1.6 / –2.8	5.0 / 9.0	+2.0 / 0	–5.0 / –7.0	5.0 / 11.0	+3.0 / 0	–5.0 / –8.0
30.09– 41.49	2.0 / 4.6	+1.6 / 0	–2.0 / –3.0	2.0 / 6.1	+2.5 / 0	–2.0 / –3.6	6.0 / 11.0	+2.5 / 0	–6.0 / –8.5	6.0 / 14.0	+4.0 / 0	–6.0 / –10.0
41.49– 56.19	2.5 / 5.7	+2.0 / 0	–2.5 / –3.7	2.5 / 7.5	+3.0 / 0	–2.5 / –4.5	8.0 / 14.0	+3.0 / 0	–8.0 / –11.0	8.0 / 18.0	+5.0 / 0	–8.0 / –13.0
56.19– 76.39	3.0 / 7.1	+2.5 / 0	–3.0 / –4.6	3.0 / 9.5	+4.0 / 0	–3.0 / –5.5	10.0 / 18.0	+4.0 / 0	–10.0 / –14.0	10.0 / 22.0	+6.0 / 0	–10.0 / –16.0
76.39–100.9	4.0 / 9.0	+3.0 / 0	–4.0 / –6.0	4.0 / 12.0	+5.0 / 0	–4.0 / –7.0	12.0 / 22.0	+5.0 / 0	–12.0 / –17.0	12.0 / 28.0	+8.0 / 0	–12.0 / –20.0
100.9– 131.9	5.0 / 11.5	+4.0 / 0	–5.0 / –7.5	5.0 / 15.0	+6.0 / 0	–5.0 / –9.0	16.0 / 28.0	+6.0 / 0	–16.0 / –22.0	16.0 / 36.0	+10.0 / 0	–16.0 / –26.0
131.9– 171.9	6.0 / 14.0	+5.0 / 0	–6.0 / –9.0	6.0 / 19.0	+8.0 / 0	–6.0 / –11.0	18.0 / 34.0	+8.0 / 0	–18.0 / –26.0	18.0 / 42.0	+12.0 / 0	–18.0 / –30.0
171.9– 200	8.0 / 18.0	+6.0 / 0	–8.0 / –12.0	8.0 / 22.0	+10.0 / 0	–8.0 / –12.0	22.0 / 42.0	+10.0 / ...	–22.0 / –32.0	22.0 / 54.0	+16.0 / 0	–22.0 / –38.0

Fig. 56.   * In thousandths of inches

The limits for the shaft are subtracted from the basic size, as follows:

    5.2500—0.0006=5.2494 (largest shaft)

    5.2500—0.0013=5.2487 (smallest shaft)

As we have said, a tolerance indicates the extent to which a certain dimension may vary. It may be written in any of the three ways shown in Fig. 57. The tolerance dimension, with the appropriate plus or minus sign before it, appears after the over-all dimension number. Sometimes the tolerance number appears below

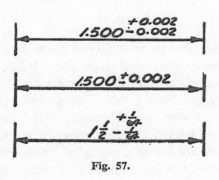

Fig. 57.

the over-all dimension, with the dimension line between. See Fig. 58. The tolerance dimension may be expressed in decimals or in fractions, but not both ways in the same drawing.

A tolerance may allow a demension to vary in one direction only, as shown in Fig. 58A. The —.002 tells us that the dimension can be .002 smaller than 1.878 but not larger.

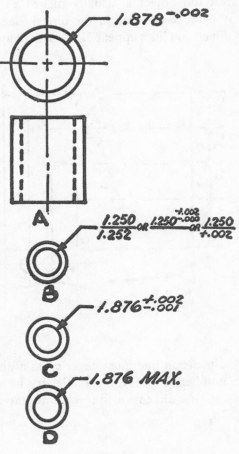

Fig. 58.

If a tolerance allows a dimension to vary in both directions, the dimension 1.876 in Fig. 58C, for example, can be increased by .002 or decreased by .001. If the tolerance allowed is the **same** in both directions, it can be written in any of the three ways shown in Fig. 57. The dimension 1.500, for example, can be **increased** or **decreased** by .002.

If many tolerances are given in one object, the person making the object could become confused and add up the tolerances, which, of course, is incorrect procedure. To avoid this, several tolerances are dimensioned as shown in Fig. 59.

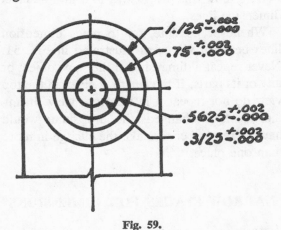

Fig. 59.

Instead of giving the tolerance, the draftsman sometimes indicates the whole minimum or maximum dimension allowed. This is called a limit system. The difference between the maximum and minimum dimensions is the tolerance. Both size and location limit dimensions are denoted by placing the high limit above the low limit. This is shown in Fig. 58B.

If there is only one limit, then only one limit dimension need be given. This is indicated in Fig. 58D. The same figure shows that the abbreviations MAX. and MIN. can be placed after the limit dimensions.

The preceding discussion is applicable to angle tolerances with the important exception that this type of tolerance is usually expressed in degrees, minutes, and seconds, or in decimals.

There is a great deal of information available on the general topic of writing tolerance dimensions. The reader might consult the standard Manual ASA Y. 14.5, published by the American Society of Mechanical Engineers.

DIMENSION LINES

Remember to allow space for extension and dimension lines when placing the three views on paper. Dimension lines must appear on the view which shows the true length of the part being dimensioned. Some views will show the object foreshortened, or will contain hidden lines. Do not put dimensions at these points. Fig. 50 shows the proper way to draw and label dimension lines.

Whenever possible, try to place dimension lines between views, as illustrated in Fig. 51. Never repeat a dimension of the same object or any of its parts. If the length is given in the top view do not repeat it in the front view. If any dimension is changed, the draftsman would have to remember to make the change in more than one place.

NARROW SPACES FOR DIMENSIONS

When the space between extension lines is too narrow for writing the dimension number, any of the methods illustrated in Fig. 60 may be used. Furthermore, when several dimension lines must be drawn near each other, the posi-

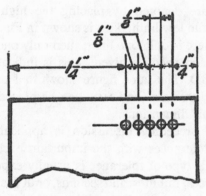

Fig. 60.

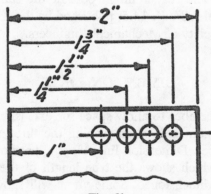

Fig. 61.

tion of the dimension numbers should be staggered, as illustrated in Fig. 61.

SPACING DIMENSION LINES

Fig. 62 shows the proper way to space dimension lines. The dimension line that appears nearest the object is usually placed ½ inch away from the outline of the object. Succeeding dimension lines appear ⅜ of an inch apart.

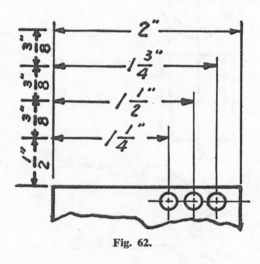

Fig. 62.

A dimension line must never cross a dimension number. Whenever possible, try to avoid crossing one dimension line with another. See Fig. 50.

LEADERS

Sometimes it is difficult to write the dimension number where it properly belongs. The space may be too small, or other figures and notes are in the way. In such cases, lines called **leaders,** illustrated in Fig. 63, are used. Leaders appear as straight lines with arrows on them pointing to the line or curve being dimensioned. The leader goes off at an angle from the line or curve and then has a short horizontal line which goes toward the middle of the dimension number. Leaders to circles and arcs should point

Fig. 63.

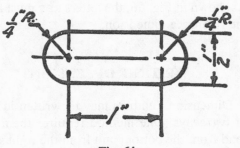

Fig. 64.

toward the center of the circle or arc. This is illustrated in Figs. 63 and 64.

Leaders should never be crossed nor should long leaders be used. They should not go in a horizontal or vertical direction nor should they be parallel to dimension, extension, or section lines. Avoid small angles in leaders.

DIMENSIONING ANGLES

To show the size of an angle, draw an arc from one leg of the angle to the other. To draw the arc, use a point at which the two legs of the angle meet as a center. The arc must be far enough away from the center so that there is enough room to write the size of the arc. Angle sizes are written in degrees, minutes, and seconds, as shown in Fig. 65. Do not use dashes.

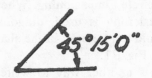

Fig. 65.

OVER-ALL DIMENSIONS

Usually one dimension which shows the whole width, length, or depth of an object is drawn. This is known as the over-all dimension. Very frequently, as shown in Fig. 66, it is necessary to draw several dimensions between the over-all dimension. When this situation occurs, the over-all dimension is always drawn farthest out from the object. One of the shorter dimensions should then be eliminated, as shown in Fig. 67. If the object being dimensioned has circular ends, as in Fig. 64, the over-all dimension is not given.

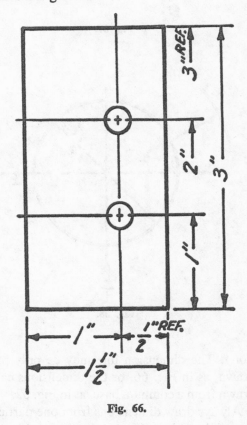

Fig. 66.

DIMENSIONS FROM A COMMON BASE

Sometimes many dimensions are used in one drawing and the available space is insufficient to permit drawing in all the necessary dimension lines. One way of overcoming this problem is shown in Fig. 68 (see the diameter dimen-

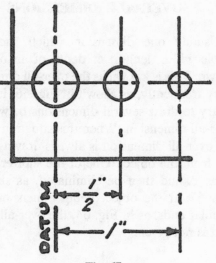

Fig. 67.

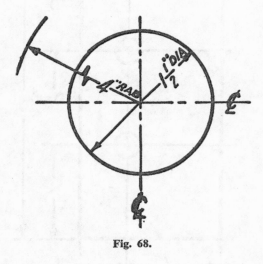

Fig. 68.

sion). The dimension line may be only partly drawn, as in Fig. 66, or the dimensions can be drawn from a common base, as in Fig. 62.

Always draw dimensions from one particular line of the object. Choose the line of the object which will allow you to draw the dimensions most legibly. See Fig. 50.

DIMENSIONS FROM DATUM

A datum refers to a part of an object about which there is no question. The part is known to be absolutely correct as to size or location.

Dimensions can, therefore, be drawn with the assurance that all the dimensions will be correct. Fig. 67 illustrates this. Before a part is accepted as a datum, you must be sure that it is absolutely correct.

REFERENCE DIMENSIONS

For information and to make some calculations easier, a reference dimension is included. As shown in Fig. 66, the letters REF must follow the reference dimension.

WRITING DIMENSIONS

Dimension numbers may be written in either of two ways. One method requires them to be read from the bottom and from the right side of the drawing. This is shown in Fig. 69. The numbers follow the direction of the dimension line.

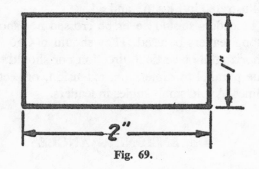

Fig. 69.

When the direction line appears horizontally, the number appears the same way. The vertical dimension lines contain dimensions written so they can be read from the right edge of the paper. If a dimension line must be drawn diagonally, the dimension number must be written in the same diagonal direction, as illustrated in Fig. 70. Diagonal dimension lines appear most often in **circle dimensioning.** When this system of dimensioning is used, diagonal dimension lines should not be used in the shaded areas, as shown in Fig. 71.

There is another way to write dimensions. Instead of writing dimensions to be read from the bottom and from the right side of the drawing, write the dimension number to be read only

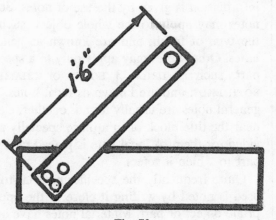

Fig. 70.

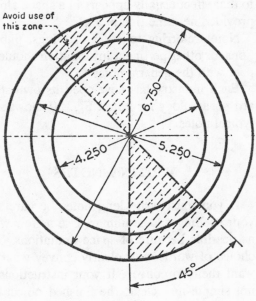

Avoid use of
this zone

6.750

4.250

5.250

45°

Fig. 71.

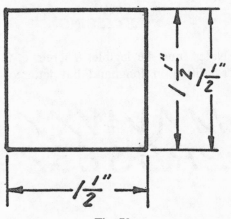

Fig. 72.

from the bottom of the drawing. Fig. 72 illustrates both methods.

Circle dimensions can be written in any position when dimensions are to be read from the bottom only.

Fig. 72 shows the proper way to write fractions no matter which method is used.

DIMENSIONING CIRCLES AND CURVES

Circles. Circles represent shafts or cylinders. In order to show the size of a shaft or cylinder the radius or the diameter is given. The radius, of course, is the distance from the center to the circle itself. The distance from one point on the circle through the center to another point on the circle in one straight line is called the diameter. The radius and diameter of a circle are shown in Fig. 68.

The dimension number should be followed by the letter "D" or by "DIA." when the **diameter** is given as a dimension. The letter "R" or "RAD." should be used when the **radius** is given. The broken dimension line goes from the center of the circle to the edge. An arrow is placed at the end of the dimension line that touches the circle. The dimension number is written on the broken dimension line. Figs. 68 and 73 show the designations for diameter and radius, the arrow, the dimension line from the center of the circle, and the dimension number on the broken line. You may sometimes see the symbol

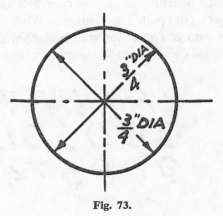

Fig. 73.

ø which stands for diameter. When very small circles are dimensioned and there is no room for the dimension number or dimension line, the method shown in Fig. 59 is used.

Center Lines. When drawing circles, lines showing the center are always included. These lines, called center lines, are drawn horizontally and vertically. This point is illustrated in Fig. 73. If there is more than one circle in the drawing, the dimensions are written from center line to center line, as shown in Fig. 67. Dimension lines, as Fig. 67 shows, are drawn to any important line of the object. The symbol ℄ may occur in some drawings. The symbol stands for center line and is written with the *C* through the center line outside of the circle, with the letter *L* through the letter C, as shown in Fig. 68.

Circles should be dimensioned on the view in which they appear as solid lines, not as dotted lines. See Fig. 50. In fact, dimensions should never be drawn from dotted or hidden lines.

Curves. Curves must be considered as parts of circles. They are dimensioned in exactly the same way as circles are dimensioned. Sometimes, however, a curve is part of a very large circle. The center of that circle may even be off the paper and therefore the dimension line is broken, as shown in Fig. 68, to indicate the size of the curve.

NOTES

Objects cannot be constructed merely by reading the dimensions. In many instances specific instructions must be given. You might want a hole drilled, an edge rounded, the whole object gold-plated or painted. What kind of brick do you want for your house? What kind of wood do you want for your hi-fi cabinet? This information is given by the use of notes. Some notes may apply to the whole object, such as the type of finish, and are known as general notes. Other notes may apply only to a specific part, such as DRILL 3 HOLES or CHAMFER BOTH ENDS, and are known as local notes. All general notes are usually found together, either near the title block or in a place specially provided for them. A local note is placed near the part to which it refers.

Quite frequently the specific material to be used is noted by writing it close to the appropriate object or part. Material notes have been standardized for many industries. If the entire object must be made of one material, the note to that effect usually appears in a space already provided near the title block.

Notes referring to other drawings, publications, or other pertinent information should appear with the general notes.

Each industry has developed its own standard phraseology for notes. Fig. 50 shows some typical notes.

WRITING NOTES

If you must write long notes be careful of your spelling and grammar. Use as few words as possible because of space limitations. Your choice of words must clearly convey what you want them to convey. If your instructions are not simple and clear, the finished object may not come out the way you expected it to.

LETTERING

Notes must be legible. Therefore, they are not written in longhand but lettered. Many

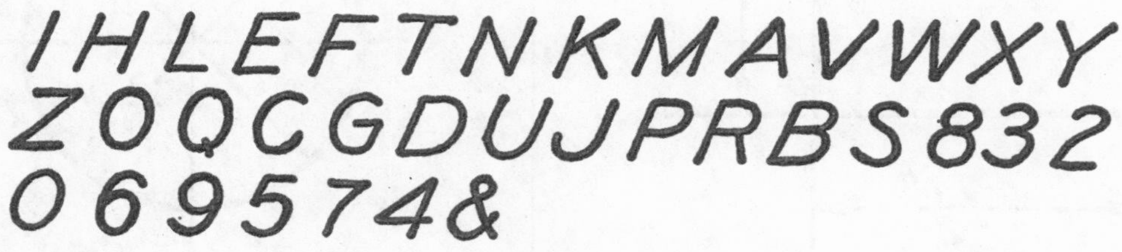

IHLEFTNKMAVWXY
ZOQCGDUJPRBS832
069574&

Fig. 74.

types of lettering have been designed. For simplicity and uniformity, the single-stroke block letter is used. **Single stroke** means the same thickness of stroke throughout the letter. The letter may be made in more than one stroke. Fig. 74 illustrates single-stroke block letters.

Except for architectural drawings, the block letter appears most frequently. Architectural drawings may require taller and narrower letters, known generally as Compressed Old Roman.

DRAWING THE LETTERS

The height of the letters depends upon the use to which they are put. Letters may be vertical or slanted. The slant is generally 67½° or in the ratio of two to five, as shown in Fig. 75. A lettering guide or triangle will provide this angle quite easily.

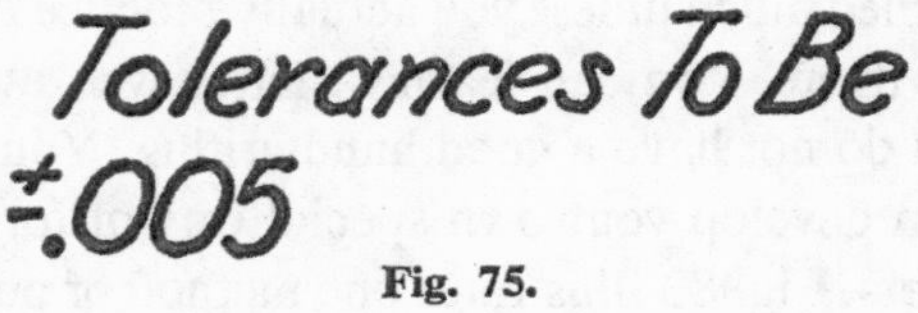

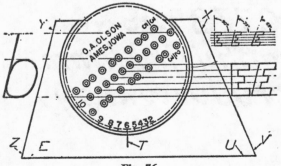

Tolerances To Be ±.005

Fig. 75.

LETTERING GUIDES

The height of letters should always appear uniform. To achieve this easily, use a lettering guide or a lettering triangle. Fig. 76 shows the Ames lettering guide. Each number from two to ten on the bottom represents 1/32 of an inch. If you want a letter ¼ of an inch high, turn the circle of the guide around until the eight matches the notch underneath the number. In

this case, 8/32 of an inch equals ¼ of an inch. Notice the three rows of holes. In the row on the left the second and fourth holes will give letters ¼ of an inch high. Put a 6H pencil with a sharp point in the second hole and push the guide along the T square. Then put the pencil in the fourth hole and push the guide back again. The third hole will give you a line 3/5 of the whole height of ¼ of an inch. The entire height could be used for capital letters, and the 3/5 height for lower-case letters. Place the first hole on the last line when you use up all the holes in the row.

The row of holes on the right, with the fraction ⅔ printed near it, is used in exactly the same way. The third hole will provide a line ⅔ of the entire height, instead of 3/5 of the entire height. The row in the center has holes ⅛ of an inch apart. It can be used for drawing letters up to a height of 1½ inches.

Proportion varies with the letter. As shown in Fig. 77, the letters C, G, O, and Q fill a square block. Fig. 78 shows that the letters H, M, N, and W also fill a square block. The letters B, D, K, P, R, X, and Z, as shown in Fig. 79, fill about 5/6 of a block. Fig. 80 illustrates that the letters A, E, F, J, L, S, T, U, and V fill about ½ of a block. Serifs or small lines at the ends of letters should not be used. Fig. 81 demonstrates the improper use of serifs.

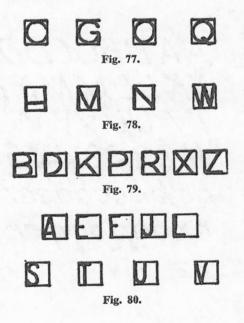

Fig. 77.

Fig. 78.

Fig. 79.

Fig. 80.

Fig. 76.

AWJKPMU

Fig. 81.

When letters are used separately to number parts, the letters l (el) and o (oh) should be omitted because they might easily be confused with the numbers one and zero.

THE LETTERING TRIANGLE

The Braddock-Rowe lettering triangle, shown in Fig. 82, is used in much the same way as the lettering guide. The large numbers—three, five, seven, and four, six, eight—refer to thirty-seconds of an inch. When the odd numbers are used, the triangle is placed on the T square so that these numbers are read right side up. When the even numbers are used, the triangle is turned around so that the long side or *hypotenuse* is on the T square. The decimals, going up toward the left, indicate the height between holes in the same row. One-quarter of an inch equals .25 of an inch. The distance between holes in each of the three rows slanting to the left is ¼ of an inch or .25 of an inch.

To get the height of 5/32 of an inch, use the first hole in the left-hand row and the first hole in the right-hand row. Use the middle row for the height of lower-case letters.

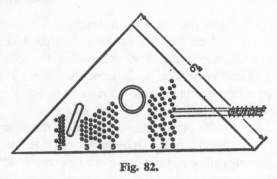

Fig. 82.

The height of all letters, other than those used in titles, should be uniform throughout the drawing. The title letters should be higher than the other letters.

PRACTICE THE LETTERS

No matter how much you read about drawing letters, you will find that you have not really learned much unless you actually practice drawing them. You can learn to letter well even if you do not have a good handwriting. You may even develop your own special way of forming letters. Fig. 83 illustrates one method of putting the strokes on paper. After you have had enough practice with an H or F pencil, try working with a 2H pencil. Then go over the pencil lines with a lettering pen.

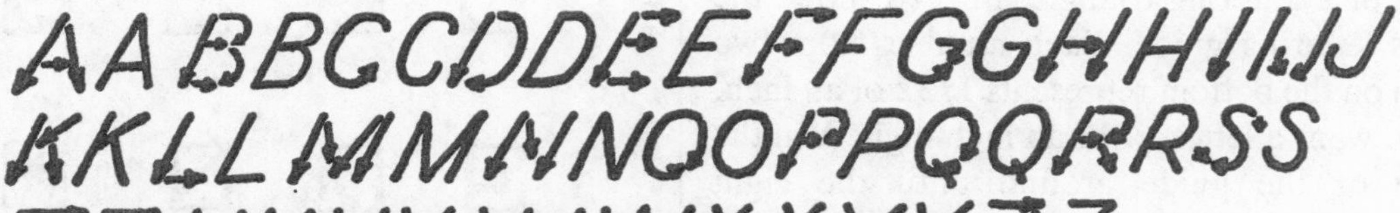

AABBCCDDEEFFGGHHIIJJ
KKLLMMNNOOPPQQRRSS
TTUUVVWWXXYYZZ
II2233445566778899&&
aabbccddeeffgghhiijjkkllmm
nnooppqqrrssttuuvvwwxxyy
zz

Fig. 83.

CHAPTER FOUR

USING THE SCALE

Scales, discussed in Chapter One, are used for drawing a large object on a smaller sheet of paper. All kinds of scales are used in the same way. Fig. 11, Chapter One, illustrates the architect's scale. Suppose you decide to make an orthographic projection of a box which is 5 feet 6 inches deep, 6 feet wide, and 4 feet 3½ inches high. See Fig. 84. (Figs. 25B and 25C,

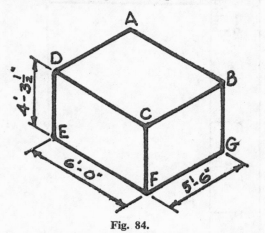

Fig. 84.

Chapter Two, illustrate the relative position of length, width, and depth.) You decide to draw the box in ¾-inch scale. This means that every 12-inch dimension of the box will be represented by a line ¾ of an inch long on paper. Start by drawing a light horizontal line of any length with a T square and a line perpendicular to it with a triangle. These lines will be part of

the top view of the box. Place the ¾-inch scale along the horizontal line with its zero mark at the point *A*, as shown in Fig. 85. The numbers to the right of the zero on the scale stand for feet. Mark off very lightly the 6-foot dimension which gives you point *B*.

The depth of the box, represented by the vertical line in Fig. 86, measures 5 feet 6 inches. The marking to the left of the zero on the ¾-inch scale represents inches and half-inches of 1 foot. The 6-inch mark would, therefore, appear in the center. Place the scale on the vertical line so that this 6-inch mark is at point *A*, as in Fig. 86. Without moving the scale, mark off the 5-foot distance which appears to the right of the zero mark. You now have the 5 feet 6 inches for the depth of the box, which is line *AD*. Draw a horizontal line at point *D*, and a vertical line at point *B*. You now have the completed top view of the box, *ABCD*, drawn to a ¾-inch scale, as shown in Fig. 87. The draftsman uses the scale only for measuring and never for drawing.

By projecting the vertical lines *AD* and *BC* down from the top view, you can draw the front view. The procedure remains the same as for the top view. Then obtain the 4 feet 3½ inches on the vertical line in Fig. 88. You find the 3½-inch mark by looking to the left of the zero.

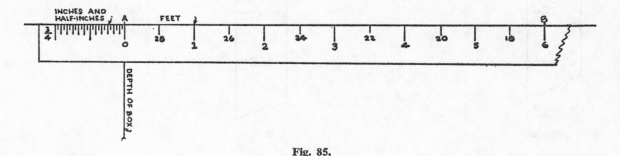

Fig. 85.

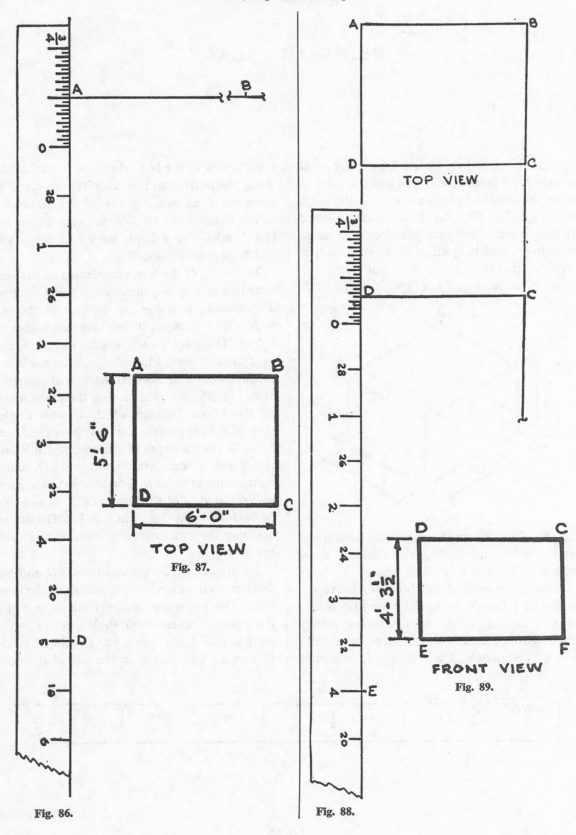

Fig. 86.

Fig. 87.

TOP VIEW

TOP VIEW

FRONT VIEW

Fig. 88.

Fig. 89.

Place the scale so that the 3½-inch mark is at point *D*. Without moving the scale, mark off the 4-foot dimension, which is to the right of the zero, point *E*. You now have line *DE*, which represents 4 feet 3½ inches. Draw the other two lines, *EF* and *CF*, to complete the front view shown in Fig. 89.

The side view requires exactly the same procedure. Extend lines *DC* and *EF*. Scale and draw the lines to their proper dimensions. Fig. 90 shows the three completed views drawn to scale.

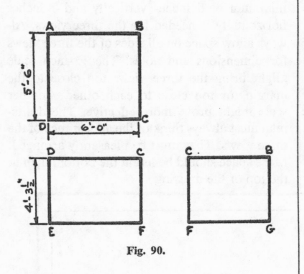

Fig. 90.

THE ⅜- AND ¾-INCH SCALES ON THE ARCHITECT'S SCALE

The length of the architect's scale is usually 12 inches, but some of them are 6 inches long. The longer one will show *14* feet on the ¾-inch scale. Just beyond the number 14 you will see a smaller-sized zero. This is part of the ⅜-inch scale. You then see a smaller number 2, which is the measurement of the second foot of the ⅜-inch scale. Number 13 stands for the third foot and so on. This is shown in Fig. 11, Chapter One. Fig. 85 shows the latter part of the ⅜-inch scale—18 to 28 feet. The 26-foot mark is also 1'-6" on the ¾-inch scale.

OTHER SCALES ON THE ARCHITECT'S SCALE

All of the other scales on the architect's scale are used the same way as the ⅜- and ¾-inch scales. The larger scales, such as the 1½-inch and 3-inch scales, are often used to draw details which contain small parts that would be difficult to read on smaller scales.

DRAWING CIRCLES AND ARCS TO SCALE

Circles and arcs are also drawn to scale. Place the needle of the compass on the zero point of the scale and extend the compass leg until the pencil point reaches the desired dimension. If the circle has a radius of feet and inches, say 1 foot 3 inches, place the needle point of the compass on the 3-inch mark, and extend the pencil leg of the compass to the 1-foot mark. See Fig. 91.

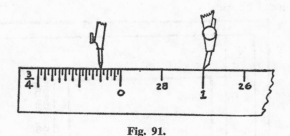

Fig. 91.

NOTING THE SCALE ON THE DRAWING

The scale used on the drawing generally appears in the title box. If details drawn to a larger scale appear on the same drawing sheet as the main object, then both scales may be noted in the title box. When more than two scales are used in the same drawing, you can write the main-object scale as ¾"=1'-0", EXCEPT AS NOTED, in the title box. The scale of each detail appears under each detail title.

The type of scale that is being used determines the form of noting the scale. Since the architect's scale shows inches or parts of inches in relation to a foot, the draftsman writes, for instance, ½"=1'-0". The mechanical engineer's scale shows inches or parts of an inch in relation to 1 inch. The draftsman, therefore, writes ½"=1". The first number of the equation, the ½", refers to the **drawing.** The other number, the 1'-0" or the 1", refers to the **object** being drawn. Therefore, ½"=1'-0" means that ½ inch on the drawing equals 1 foot on the object.

You may sometimes hear the words "quarter size." They do not mean quarter scale. **Quarter size** means ¼"=1" on the mechanical engineer's scale, while **quarter scale** means ¼"= 1'-0" on the architect's scale.

DETERMINING THE PROPER SCALE

The draftsman will often have the problem of determining the proper scale to use. The two factors he must take into consideration are the size of the object and the space allowed him on his drawing paper. The draftsman must also leave enough room between the views to draw the dimension lines and numbers, and to insert local notes.

The draftsman must first get the over-all dimensions of the object. Then he must measure the drawing space allotted on the paper, using an ordinary ruler. He must remember that three views are required.

If, for instance, the object is 10 feet wide and 4 feet deep, and a ½-inch scale is used, the front view will take up 5 inches horizontally and the side view will take up 2 inches horizontally. Study Fig. 92. Still working with Fig. 92, if the same object is 6 feet high, the front view will take up 3 inches vertically, and the top view will take up 2 inches vertically. Therefore, a minimum of 5 inches vertically and 7 inches horizontally is needed for the three views. Always allow space on all sides of the three views for dimensions and notes. The ½-inch scale might bring the three views too close to the margin, or too close to each other. Another scale might prove more effective. The draftsman must always think of the appearance of the three views. They must be pleasantly arranged. More space should be left at the bottom than at the top of the drawing.

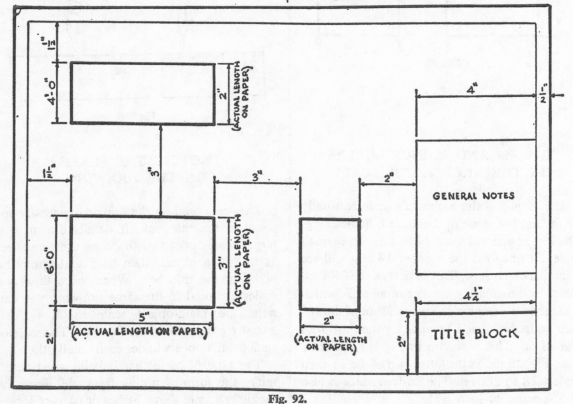

Fig. 92.

CHAPTER FIVE

SECTIONS

The three views and the auxiliary views to be studied in Chapter Six show the outside shapes of objects. Most objects drawn in orthographic projection have many interior details. Hidden lines cannot be used to show these details because the drawing would become too complicated. How do you show the plumbing or wall construction of a house or how to put together the sides and top of a chest of drawers? The answer is by simply cutting the object into parts and drawing what is seen on the inside. Of course, certain rules must be followed in cutting and drawing, so that everyone will understand what was done. In order to draw a section the draftsman should first have three views, or enough information to know what the interior of the object looks like.

SECTIONAL VIEW

That part of an object which is cut and drawn is called a **sectional view** or, more commonly, **a section.** The **section** must always be made in **elevation.** No other view is used. In most cases, hidden lines are not shown in the sectional view.

FULL SECTIONS

If the object is cut right through as though with a knife, and the exposed inside surfaces are drawn in elevation, the result is a full section, as shown in Fig. 93. The object does not have to be cut through in one straight line. The cutting could be off-set, as in Fig. 94. In general, the object should be cut to show the interior most clearly. The cutting angles should be kept at 90°.

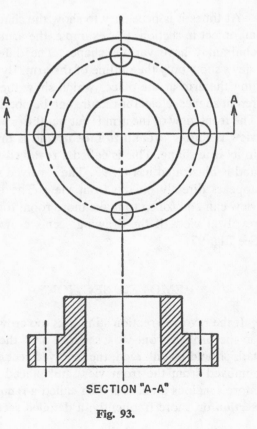

SECTION "A-A"

Fig. 93.

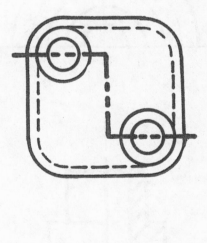

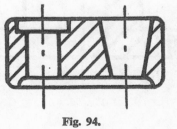

Fig. 94.

HALF SECTIONS

In many symmetrical objects the details shown in full section are the same on either side of the center line. When this occurs the draftsman, so as to save time and to make the drawing simpler and clearer, will draw a **half section,** illustrated in Fig. 95, instead of a full section. As its name indicates, a half section shows half of a full section. The other half of the object appears as the regular front view. Hidden lines are not included in either part of the half-section view.

BROKEN-OUT SECTIONS

When only a small part of an object requires a sectional view, the draftsman simply draws

the complete front view. With a broken line, drawn freehand, he marks off the part to be shown in section and draws that part only in section. This is called a **broken-out section** and is illustrated in Fig. 96.

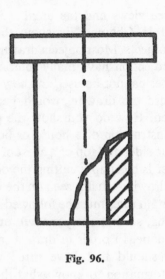

Fig. 96.

REVOLVED SECTIONS

At times it is necessary to show the shape of an object in section. For example, the arm of a chair may have various shapes. The different views show only the outline of the arm. By cutting the arm at the places which show the different shapes, a clearer picture can be obtained. The front view of the arm is cut but **the exposed view is turned around to face us** and is drawn in its **true shape.** This is called a **revolved view** and is illustrated in Fig. 97. The revolved view appears directly on the front view. The front view can be broken to make more room for the revolved view, if the drawing seems crowded. See Fig. 97.

REMOVED SECTIONS

If the revolved section still looks too crowded on the broken front-view section, or if the details should be enlarged, the revolved section is removed from the front view and placed in a more spacious area. It is then called a **removed section** or, more frequently, a **detailed section.**

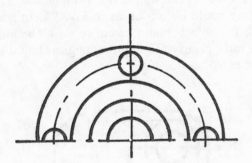

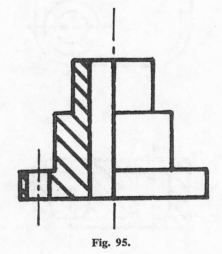

Fig. 95.

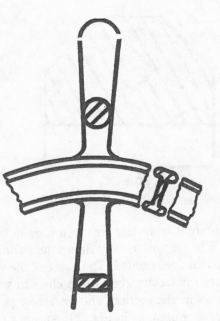

Fig. 97.

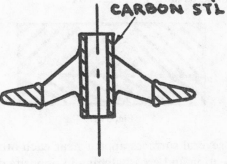

Fig. 99.

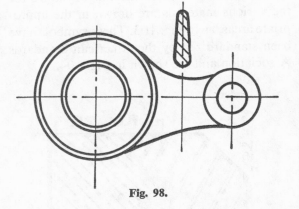

Fig. 98.

Fig. 98 shows a removed section. For simplicity and clarity all detailed sections of one object are sometimes drawn on separate sheets of paper.

FORESHORTENED SECTIONS

Parts of some objects will appear foreshortened in a full-section view. Fig. 99 shows a **foreshortened section.** For reasons of clarity, the foreshortened part is revolved so that the drawing is symmetrical. Drawings of standard industrial pieces, such as flanges, ribbed pulleys, or wheels, always appear without any foreshortening.

CROSSHATCHING AND SYMBOLS

In order to distinguish between a section and an ordinary elevation, all surfaces which come in contact with the cutting knife are crosshatched, as shown in Fig. 94. Those surfaces, such as the hole, which the knife does not touch are not crosshatched. See Fig. 93. The lines are made with a 45° triangle and are spaced evenly, about ⅛ of an inch apart. Use your eyes to space the lines. Measuring off equal spaces with a ruler would require too much time. If the 45° angle at which the crosshatch lines are drawn coincides with the outline of the object, use another angle for the crosshatch lines, as shown in Fig. 100.

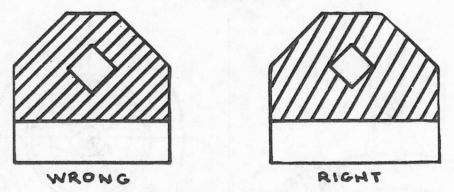

Fig. 100.   Drawing Lines at Proper Angle.

Surfaces appearing next to each other in the section may not be next to each other in the actual object. To show that surfaces are not in the same plane, crosshatch them in opposite directions, as shown in Fig. 101.

Fig. 101.

If several surfaces appear near each other in the section and crosshatching in opposite directions does not help, the closeness of the crosshatched lines can be varied. See Fig. 102.

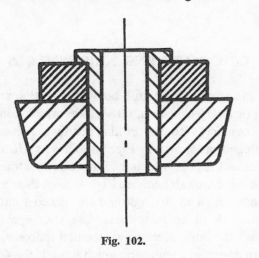

Fig. 102.

When the same surface appears in different parts of the section, the crosshatching is drawn in the same direction in each part, as shown in Fig. 99.

Different materials are often used in one object. The section usually shows the various materials that are needed. The name of the specific material is frequently written close to where it appears in the section, and an arrow points to the proper area, as in Fig. 99. Another method of differentiating materials involves the use of symbols. Instead of crosshatching, symbols for the various materials are drawn in the appropriate areas, as in Fig. 103. These symbols have been standardized by the American Standards Association and are shown in Fig. 104.

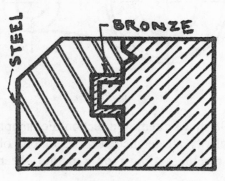

Fig. 103.

THIN SECTIONS

The draftsman is frequently required to draw objects which are too thin for crosshatching. Such objects include structural beams, thin

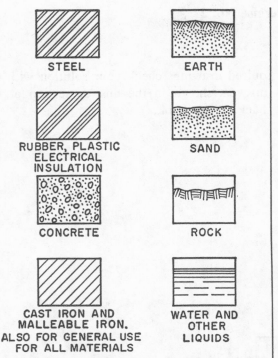

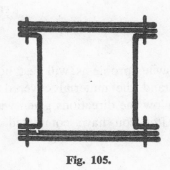

Fig. 105.

Fig. 104. **Symbols for Section Lining.**

wood pieces, gaskets, etc. The section areas are not crosshatched, but are drawn as solid lines, somewhat heavier than object lines. See Fig. 105.

CUTTING-PLANE LOCATION

How do you show the shape of the section taken or the direction in which the exposed inside surfaces face? Fig. 93 illustrates the use of lines and arrows. The lines, which show the place or places that were cut, are not drawn throughout the object. The more lines that appear on a drawing, the more confused the drawing may become. Therefore, no unnecessary lines should appear on a drawing. To show a full section that is taken in one straight line, draw short heavy lines on each side of the plan view. These lines must lie on one straight line,

exactly where the object is being cut. Then draw lines perpendicular to these lines, putting arrows, pointing in the direction we are facing, on the ends. The lines, with the arrows on them, must include only the area to be shown in section. If a full section along a center line is taken, arrows and cutting lines are not needed. Remember that although the cutting lines are drawn on the plan view the section will be shown in a front view. In order to draw a section the draftsman must have the three views, or enough information to know what the interior of the object looks like.

If a full section is not taken on one straight line, then short lines are drawn on the plan view to show the path of the cut. See Fig. 94.

Cutting lines and directional arrows for removed or detailed sections must be shown. They are not required for revolved sections or broken-out sections.

Letters are used more frequently than numbers to denote sections. Letters are put near the directional arrows and are used for the title of the section. This is shown in Fig. 93.

If sections are placed on a separate sheet, the number of the sheet is put near the letter of the section to expedite location.

Practice Exercise No. 2

The following problems will test how well you understand the material covered in this chapter. Follow the directions given with each problem. After you have completed the required drawings check your solution with the answers shown in the answer section at the back of the book.

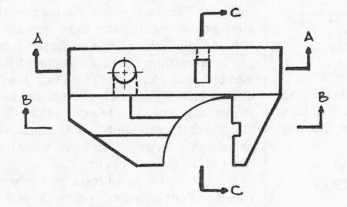

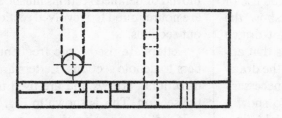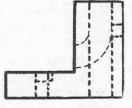

PROBLEM 1: Draw sections as shown.

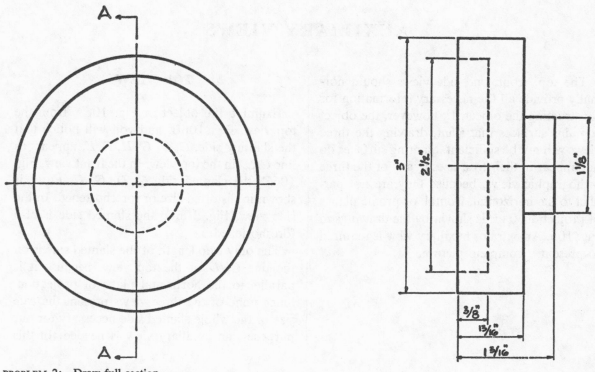

PROBLEM 2: Draw full section.

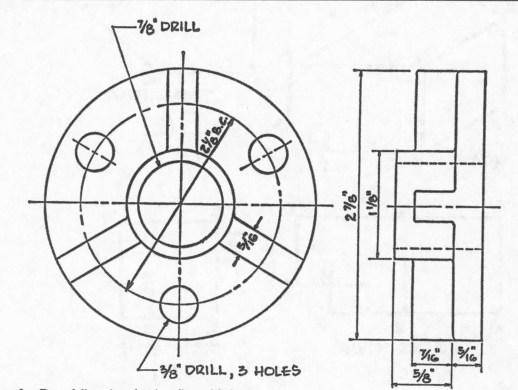

PROBLEM 3: Draw full section showing ribs and holes.

CHAPTER SIX

AUXILIARY VIEWS

The top, front, and side views should normally provide all the necessary information for constructing the object. If, however, the object contains surfaces that slant, drawing the three views will not be sufficient. Slanting surfaces do not appear in their true sizes in any of the three orthographic views, because they are not parallel to the horizontal, frontal, or profile planes of projection. This is shown by the drawings in Fig. 106. An extra or auxiliary view is required to present a complete picture.

TOP VIEW

Examine the object in Fig. 106A from the top view, Fig. 106B, and you will notice that the slanting side, *C, D, G, H, F, E*, appears as line *CHD* in the top view. In the front view, Fig. 106C, the slanted side *C, D, G, H, F, E*, is shown in distorted size or foreshortened. In the side view, Fig. 106D, the slanted side is also foreshortened.

The only true length of the slanted side is on the line *CHD* of the top view, because it is parallel to the horizontal plane of projection. Since none of the three views present the true size of the whole slanted side necessary for our purposes, an auxiliary view is needed. In this

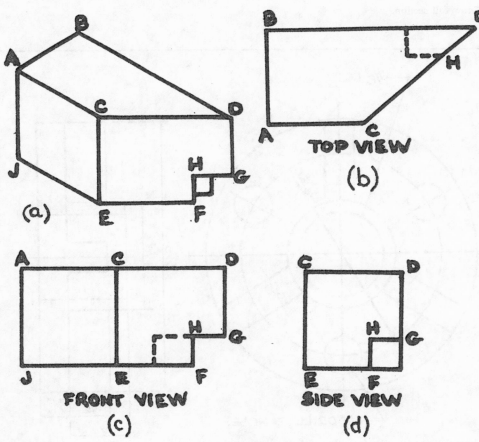

Fig. 106.

instance the auxiliary view is drawn from the top view because the true length of one edge of the slanting side is known for this view. An auxiliary view is always drawn from an orthographic view which shows one edge in its true length.

Imagine that the line *CHD* is on hinges. Swing the slanted side, *C, D, G, H, F, E,* on the hinges until the side is flat with the top surface, *A, B, D, C.* Pull *C, D, G, H, F, E* away from the top surface and you have the true size of the slanted surface. This is shown in Fig. 107.

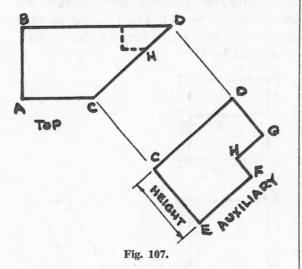

Fig. 107.

To draw this auxiliary view, still working with Fig. 107, start by making the three normal views, top, front, and side. See which view has

a true length. In this problem the top view has it. Draw projection lines perpendicular to *CD*. To these perpendicular lines draw *CD* again. See Fig. 108.

From the front view in Fig. 107 measure the true height of the slanted surface, *C to E.* Put this height on the projection line from *C.* Now measure the true distance from *D to G* on the front view and put that distance on the projection line from *D.* The true distance from *E to F* appears in the top view from *C to H.* Measure this distance and put it on the auxiliary view from *E to F.* Again the true distance from *H to G* appears only in the top view from *H to D.* Measure the distance from *H to D* in the top view and mark it on the auxiliary view from *G.* See Fig. 107.

The true vertical distance from *F to H* appears in the front view. Mark this distance on the auxiliary view. The point *H* occurs at the crossing of this vertical distance from *G to H.* By connecting the points the auxiliary view is obtained. Notice that only the slanted surface appears as the auxiliary view. Never project the other surfaces or the entire object.

FRONT VIEW

Suppose the object, Fig. 108A, has a slanted surface which appears as a slanted line in the front view, as shown in Fig. 108B. The slanted line is parallel only to the frontal plane of pro-

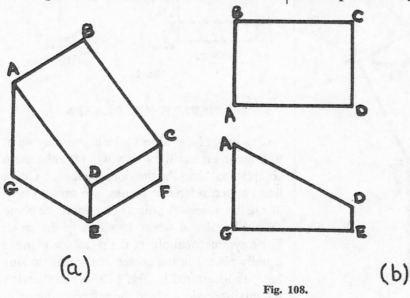

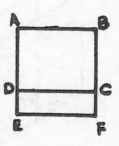

(a)   (b)

Fig. 108.

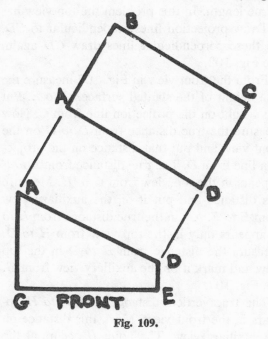

Fig. 109.

jection. The other views do not show the slanted line in its true length and therefore the auxiliary view must be drawn from the front view. This is illustrated in Fig. 109.

Draw projection lines perpendicular to *AD*. Draw *AD* again to the projection line. Lines *AB* and *DC* appear in their true lengths in the

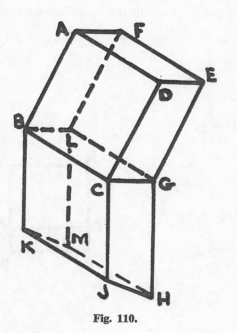

Fig. 110.

top view. Mark off the points *B* and *C* from *A* and *D*. Connect the lines and you have the auxiliary view.

SIDE VIEW

When one of the edges of the slanted surfaces is parallel to the profile plane of projection, it appears in its true length in the side view. The auxiliary view is then drawn from the side view, as in Fig. 110.

As in the other problems, the three views are drawn first, as shown in Fig. 111, and then the side view is used as the base for the auxiliary view. See Fig. 112.

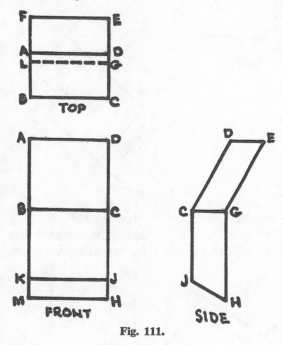

Fig. 111.

REFERENCE PLANES

The part of the object which you may want to draw in an auxiliary view will often be more complicated than the previous examples. Guide lines, called reference planes, are used to help locate the necessary points. The reference plane should be placed where it will help the most. For a symmetrical object the reference plane is usually placed in the center of the slanted surface, as illustrated in Fig. 113. When drawing the auxiliary view, place the reference plane in

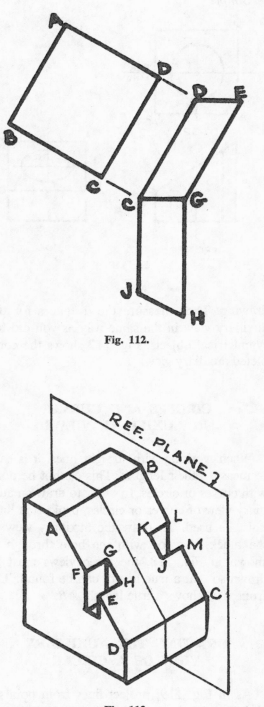

Fig. 112.

Fig. 113.

view by marking off the true distances on both sides of the reference plane. The distances from *X to A* and from *X to B* are the same as the distances from *Y to D* and *Y to C*.

Locate point *F* by measuring from point *A*, and point *L* by measuring from point *B*. From point *C* find point *M*, and from point *D* find point *E*. The distances from line *XY* of the reference plane to points *G, H, J,* and *K* are the

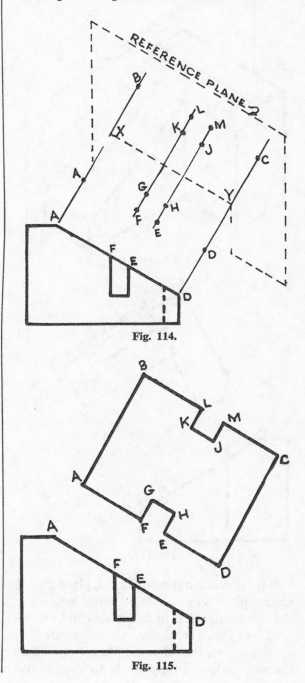

Fig. 114.

Fig. 115.

the same position it has in the perspective drawing, along the center. Then locate all the points from the reference plane.

After projecting lines from *AD* for the auxiliary view, draw a reference plane. See Fig. 114. Then locate the points of the auxiliary

same. Draw a perpendicular line from point *F* to line *XY*. Measure point *G* on this line from reference line *XY*. Find the other points, *H, J,* and *K,* by drawing perpendicular lines from points *E, M,* and *L* to line *XY.* Fig. 115 shows the completed auxiliary view.

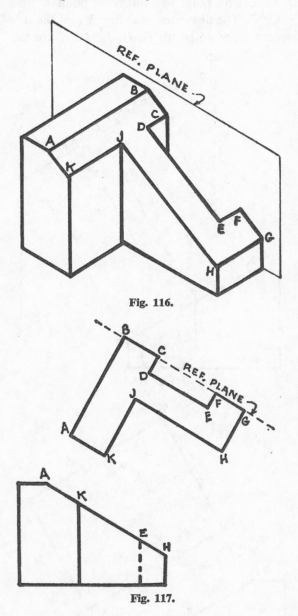

Fig. 116.

Fig. 117.

If the object is not symmetrical, place the reference plane to one side of the object. Select the side from which all the points can be measured. As Fig. 116 shows, when the auxiliary view is drawn, the reference plane is placed on the same side as it appears in the perspective

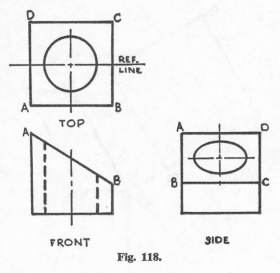

Fig. 118.

drawing. Then measure the distances for the auxiliary view in the same way as you did for symmetrical objects. Fig. 117 shows the completed auxiliary view.

CIRCLES AND CURVES IN AUXILIARY VIEWS

When working with straight lines, it is easy to measure their lengths. This cannot be done with curves or circles. In order to draw an auxiliary view of curves or circles, projection lines must be used. To draw the auxiliary view of the block of wood with the hole through it, shown in Fig. 118, the three views must be drawn so that a true length can be found. The front view shows a true length, *AB.*

DRAWING THE AUXILIARY VIEW OF A CIRCLE

As in Fig. 119, project lines from points *S* and *T* in the top view down to the slant line in the front view. From the point where they meet on the slant line, project perpendicular lines. At any convenient place, draw a line parallel to line *ST* of the front view. This line will serve as an auxiliary reference line.

Pick any points on the circle, such as *A, B, C, D, E, F, G,* and *H.* Draw lines down from these

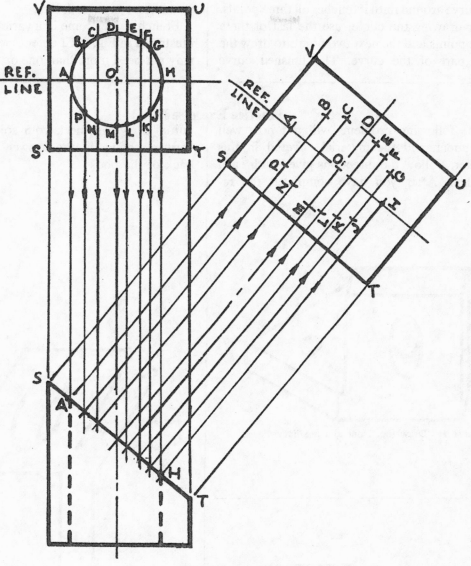

Fig. 119.

points to the slant line. Where they cross the slant line, draw perpendicular lines to cross the auxiliary reference line.

With a bow compass, measure the distance in the top view from point *O,* the center of the circle, to point *D.* Place the needle of the compass in point *O* on the auxiliary reference line and mark off the distances *OD* and *OM.* The distances *OD* and *OM* are equal because the reference line is in the center of the circle.

With the bow compass, measure the distance from the reference line on the top view to point *B.* Follow the line that goes from point *B* to the front view, then up to the auxiliary view. Put the needle of the compass at the point where this line crosses the auxiliary reference line. Mark off the distance on both sides of the reference line to get points *B* and *P.* Repeat this procedure for all the other points. Connect the points with the French curve.

USE OF THE FRENCH CURVE

The French curve should not be used to connect more than three points at a time. Move the curve around until it touches all three points. After drawing the curve, use the last of these three points and the next two points to draw the next part of the curve. The finished curve should be one smooth line.

French curves come in a variety of shapes, as illustrated in Fig. 6, Chapter One. The beginner may not need more than one or two.

Practice Exercise No. 3

The following problems will test how well you understand the material covered in this chapter. Follow the directions given with each problem. After you have completed the required drawings check your solution with the answers shown in the answer section at the back of the book.

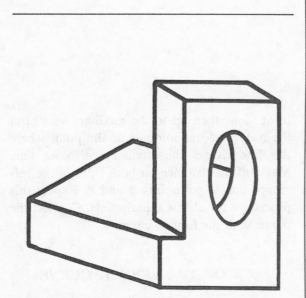

PROBLEM 1:   Draw top, front and auxiliary views.

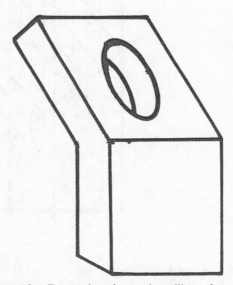

PROBLEM 3:   Draw edge view and auxiliary view.

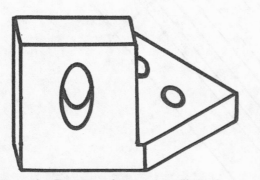

PROBLEM 2:   Draw top, front and auxiliary views.

CHAPTER SEVEN

DEVELOPMENTS

In many instances the three orthographic views and auxiliary views will not be adequate for our purposes. The object to be constructed might, for example, be made of sheet metal or have a very odd shape. A three-dimensional sheet-metal object, whenever possible, stems from one flat sheet. Several kinds of sheet-metal machines roll, bend, or fold the sheet into the desired shape. The problem is to draw a full-size pattern of the flat sheet. This type of drawing is called a **development.**

The object will be made exactly the same size as the drawing the draftsman makes. The pattern must, therefore, be drawn to the true dimensions of the object. The pattern should be drawn as if you were looking at the inside of the object.

The edges of the object must be joined in some way. Welding provides the strongest connection, but increases the cost of production. Several methods of **folding over** the edges on each other, other than welding, have been devised. The additional material must be included on the pattern so that the manufacturer will be able to do the folding over, or **seaming,** as it is called. Various sheet-metal machines perform the seaming operations for metal objects.

DEVELOPING A BOX

Fig. 120 shows a four-sided hollow box with a top and bottom. The problem is to make a drawing from which a pattern will be cut. By folding the pattern the entire box, including the top and bottom, will be constructed. The draftsman, therefore, must draw only the true lengths of all six sides. The development will not include the extra material for seaming. **All developments appear as a front view only.** First draw the three orthographic views, top, front, and side, as shown in Fig. 121. From the front view the true height of the box, E to D, the vertical distance from the bottom line, E to D in Fig. 123, and the true length of one side of the box, E to F, can be determined.

Drawing the Development. Since all developments are drawn as elevations, the front view of Fig. 121, which is also an elevation, can be used as a basis for our drawing. Fig. 122 shows the front view. Extend the line EF to the right. At any convenient point on this extended line, mark off the distance EF.

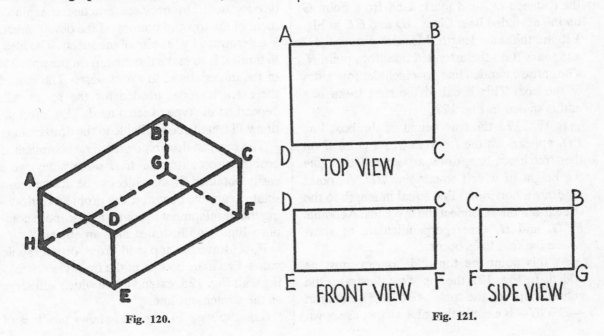

Fig. 120.

TOP VIEW

FRONT VIEW

SIDE VIEW

Fig. 121.

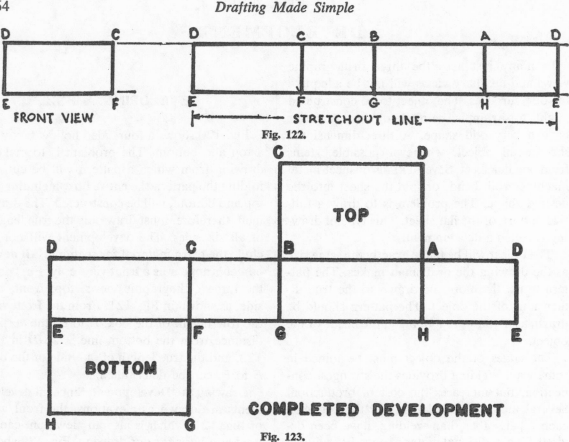

FRONT VIEW   STRETCHOUT LINE

Fig. 122.

TOP

COMPLETED DEVELOPMENT

BOTTOM

Fig. 123.

In Fig. 121 measure the distance *FG* in the side view. From point *F* on the extended line, mark off the distance to point *G*. Lines *BA* and *GH* in Fig. 120 are the same length. Measure the distance *DC* and mark it off from point *G* on the extended line. Lines *AD* and *HE* in Fig. 120 are the same length. Measure distance *AD* and mark it on the extended line from point *H*. The entire extended line now includes four sides of the box. This is called the **stretch-out line** and is shown in Fig. 122.

In Fig. 122 the true height of the box, line *ED*, appears on the front view. At point *E* on the stretch-out line, draw a perpendicular line the height of which equals line *ED*. At point *D*, draw a horizontal line equal in length to the stretch-out line. Connect the two ends. At points *F*, *G*, and *H*, draw perpendiculars to show where the fold lines occur.

At this point the top and bottom must be added. In Fig. 121 the top view illustrates the actual shape of the top—*ABCD*. The bottom —*EFGH*—is exactly the same shape, as shown in Fig. 123. To draw the top, add the sides *AD*, *DC*, and *CB* to line *BA* in Fig. 123. Add lines *FG*, *GH*, and *HE* to draw the bottom. They may be added to any of the horizontal edges of the development. In business practice the placement of the top and bottom on the development is determined by the size of the sheets. The idea, naturally, is to make the maximum possible use of the material and to avoid waste. The size of the extra material needed for the seams will depend on the type of seam used. This information will usually be available to the draftsman.

The box just drawn presented no complicated problems because the true dimensions were easily obtained. If the object to be drawn has a slanted surface, however, the problem of drawing the development becomes a little more complex. Fig. 124 illustrates such an object.

First draw the top and front orthographic views. From the bottom of the front view, working with Fig. 125, extend a line which will serve as the stretch-out line.

The top view in Fig. 125 shows the lines of

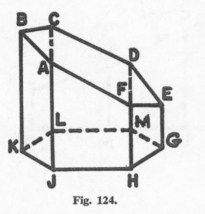

Fig. 124.

the base in their true lengths. The front view of the same figure shows the various heights of the object in their true dimensions. Put the lines of the base on the development by measuring the length of one line and marking it on the stretch-out line six times, since all six base lines are of equal length.

To show the various heights, project lines from points *B*, *A*, *F*, and *E* in the front view, across and parallel to the stretch-out line. Now draw perpendicular lines at points *G*, *H*, *J*, *K*, *L*, *M*, and *G*, as in Fig. 125. The true heights occur at the crossing of these perpendiculars with the horizontal lines from the front view. Draw lines to the cross points, *E*, *F*, *A*, *B*, *C*, *D*, and *E*.

The first line *EG* should equal the last line *EG* in length. They will meet when the pattern is folded to form the object. Use the shortest

elevation, line *EG* in Fig. 125, as the first one in the development. The shortest elevation requires the least amount of seaming, riveting, or welding, which keeps down the cost of making the object. The shortest elevation is found by examining the front view.

The top and bottom are still to be added. The bottom can be drawn along any one of the edges shown on the bottom line. Let us use the edge *H* to *J*. The edges of the bottom will be attached to the edges of the bottom line. When the sheet is folded, *JK'* will be joined to *JK*, *K'L'* to *KL*, *L'M'* to *LM*, etc. See Fig. 127. Therefore, the length of *JK'* must be the same as *JK*, and the length of *K'L'* must be the same as *KL*, etc.

Now try to locate point *K'*. Put the compass in the point *H* and measure the distance to *J*. With point *J* as the center, draw an arc as in Fig. 126. Now measure the distance from point *D* to point *F* in the top view. See Fig. 125. Mark that distance on the development from *J* in a line perpendicular to *HJ*. This gives us the distance across the bottom, *JL*. Measure the length from *K* to *L*. Mark this length from *L'* by crossing the first arc made from *J*. This crossing gives us the point *K'*. Lines *HJ* and *L'M'* are of equal length, so measure *HJ* and mark it off from *L'*. The point *G'* appears on the same line as *K'*. Draw a light line across from *K'*. Measure the distance from *L* to *M* on the *bottom line*. From *M'* mark this distance

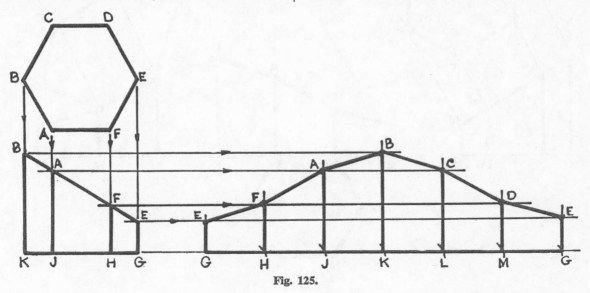

Fig. 125.

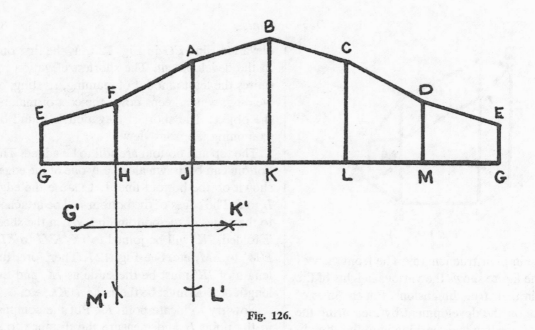

Fig. 126.

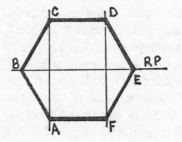

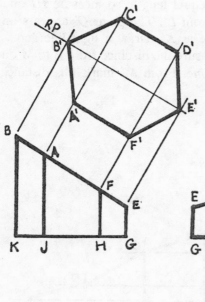

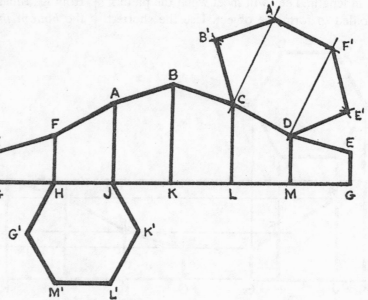

Fig. 127.

on the light line drawn from *K'*. This gives us *G'*. Complete the bottom outline by drawing a line from point *G'* to point *H*.

To draw the cover of the prism, the draftsman must first draw the auxiliary of the slanted surface. The auxiliary will give the true shape of the cover, which is the slanted surface. After drawing the auxiliary according to the method described in Chapter Six, transfer it to the development. The cover may be attached to any one of the edges of the development. At edge *CD* in Fig. 127 two perpendiculars are projected from points *C* and *D*. With the compass, take the distance *C'A'* from the auxiliary and transfer it to the development by marking off a small arc from point *C*. Since *C'A'* equals *D'F*, mark off the same distance from point *D* on the development.

With the compass, take distance *AB* on the development and make a small arc from point *A'*. Take the distance *BC* on the development and mark the distance from point *C*, crossing the small arc just made from point *A'*. The crossing of the two arcs gives us point *B'*.

Take distances *FE* and *DE* from the development and obtain point *E'* by the same method. Connecting the points will give us the true shape of the cover.

VERTICAL SLANTING SURFACES

Objects requiring development drawings will sometimes have slanting sides. The method of drawing this type of development differs from the method used for drawing an object with a slanted top. Fig. 128 illustrates the object.

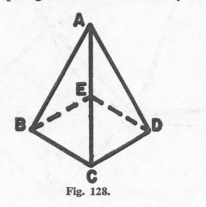

Fig. 128.

FINDING THE TRUE LENGTH OF A LINE

First draw the top and front views of the pyramid, as shown in Figs. 129 and 130. The

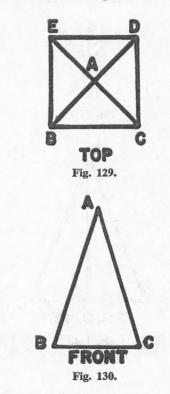

TOP
Fig. 129.

FRONT
Fig. 130.

side view need not be drawn since it would look just like the front view. Remember to use only the true bottom—*BCDE*, in the top view, as appearing in its true dimensions. The height, however, from *A* to the corners, *B, C, D,* and *E*, does not appear in its true dimension in the top view. The distance *A* to *B*, which appears in the top view as the height, becomes foreshortened, and in the front view the height *A* to *B* is really slanting in from *B* while it goes up to *A*, and therefore does not appear in its true dimension. To find the true length of *AB*, turn the object so that *AB* appears in its true length when you look directly at it. The top and front views will then appear as in Figs. 131 and 132. The object is really turned until *AB* becomes parallel with the sheet of glass to which it is projected. At this point it is suggested that you review Chapter Two, relating to Orthographic Projection.

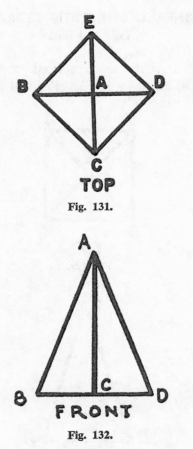

Fig. 131.

Fig. 132.

The object can be turned without our having to redraw the views. This is shown in Fig. 133. Put the compass point in *A*, in the top view, measure the distance to *B*, and describe an arc. On a line from *A*, parallel to *BC*, is point *B'*. Now project a line down from *B'* to the front view. *AB'* becomes the true length of *AB*.

Drawing the Development. When the true length of line *AB* is known, the development of the pyramid can be drawn. Use the true length of *A* to *B*, which is *AB'*, as a radius and draw an arc, as in Fig. 134.

Continuing with Fig. 134, measure the distance from *B* to *C* in the top view and mark that distance on the arc. Continue the procedure until all four sides of the bottom are marked on the arc. The bottom of the pyramid may be added to any one of the four bottom edges—lines *B'E'*, *E'D'*, *D'C'*, or *C'B'*. If the bottom of the pyramid is added to line *D'E'*, as shown in Fig. 134, get the true lengths of *D'C'* and *B'E'* in the development and draw them perpendicular to line *D'E'* since the base is a square. Complete the drawing of the bottom by drawing line *B'C'*.

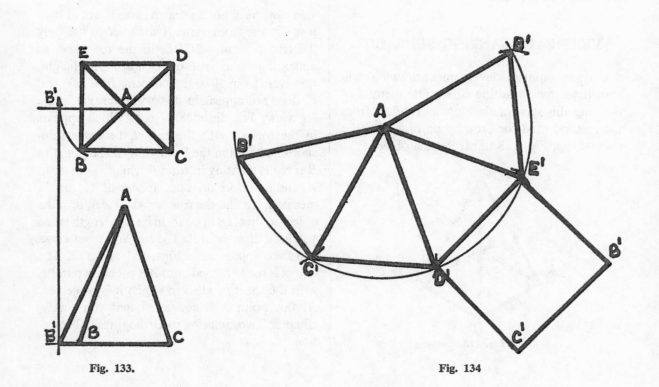

Fig. 133.   **Fig. 134**

OBJECTS WITH UNEVENLY SLANTING SIDES

The object just drawn has sides that slant toward each other at the same angle. That puts point *A* right above the center of the bottom. When we found the true length of *AB* we also found the true length of all the sides, because they are all of equal length.

To draw an object that has sides that slant toward each other at different angles, such as the object shown in Fig. 135, follow the same procedure in drawing the development that is shown in Figs. 136 and 137. Note that the lines going to the top will not be of equal length, as shown in Fig. 135. Now find the true length of the unequal lines separately. See Figs. 136 and 137.

Draw the top and front views of the figure. In the top view, Fig. 136, draw a horizontal line through point *A* parallel to side *CD*. Find the true length of each sloping side, as explained in our discussion of Fig. 133.

In Fig. 136 the line *AB'* in the front view is the true length of line *AB* in the top view. To draw the development, at any convenient spot, with point *A* as the center, mark the length of line *AB'* with a small arc.

In the top view, Fig. 136, the base *BCDE* appears as its true shape, and the four sides, *BC*, *CD*, *DE*, and *EB*, are therefore true lengths. Use these true lengths in drawing the development shown in Fig. 137. Measure line *BC* with the bow pencil and mark its length on the development as a small arc, using point *B'* as the center, as shown in Fig. 137.

The line *AC'*, in the front view, appears as the true length of line *AC*. Using point *A* in the development as a center, cross the small arc previously made from point *B'*, which gives us point *C'*.

Measure the length of line *CD* in the top view, and line *AD'* in the front view, to obtain point *D'* in the development. Follow the same procedure for getting points *E'* and *B'*. Connect all points to complete the development.

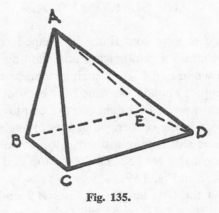

Fig. 135.

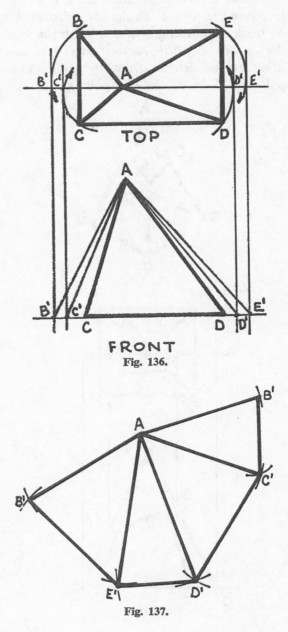

TOP

FRONT

Fig. 136.

Fig. 137.

OBJECTS WITH SLANTING TOP AND SLANTING SIDES

Objects may sometimes be shaped so that they contain a slanting top and slanting sides, as shown in Fig. 138. In such instances, use the techniques already explained for each of these circumstances. In other words, employ both methods in the one drawing. First, imagine the figure as still having a pointed top at point *A*, as shown in Fig. 138. Draw the top and front views, as in Fig. 139.

Find the true lengths of lines *AB* and *BC*. With distance *AB′* as a radius, describe a large arc, as in Fig. 140. Draw line *AB′*. From point *B′*, mark off four lengths of side *BC*, taken from the top view, Fig. 139. All four sides of the base are the same length. Draw light lines from points *B′*, *C′*, *D′*, *E′*, and *B′* to point *A*.

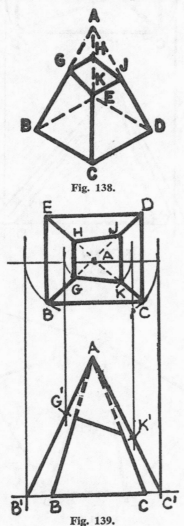

Fig. 138.

Fig. 139.

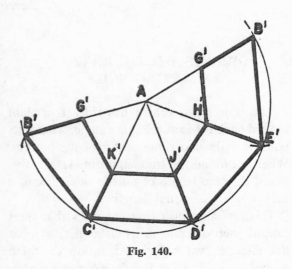

Fig. 140.

Now the sides of the slanted top must be added to the development. To find their true lengths, proceed as you did in finding the true lengths of the sloping sides. See Fig. 139. With point *A* in the top view as the center, swing an arc from point *G* on line *AB* to the line parallel to side *BC*. Then drop a perpendicular line until it crosses the true length, *AB′*, in the front view. Repeat this procedure with point *K* on line *AC*. This gives the true lengths *AG′* and *AK′*.

Measure the distance *AG′* with the bow pencil and mark it with a small arc on line *AB′* of the development. Do the same with distance *AK′*. Since the distance from point *A* to point *G* equals the distance from point *A* to point *H* and the distance from point *A* to point *K* equals the distance from point *A* to point *J*, length *AG′* can be used to locate *AH′* on line *AE′* in the development. Locate *AJ′* in a similar fashion by using distance *AK′*. Connect the points, as shown in Fig. 140, to complete the development.

CIRCULAR OBJECTS

Cylinders. Circular objects are drawn in much the same way as the objects previously described. Only true lengths may be used in drawing circular objects. To obtain the development of the cylinder in Fig. 141, it must really be cut open from top to bottom and rolled out.

In order to draw the cylinder to scale the *true height* and *true width* must be known. The ver-

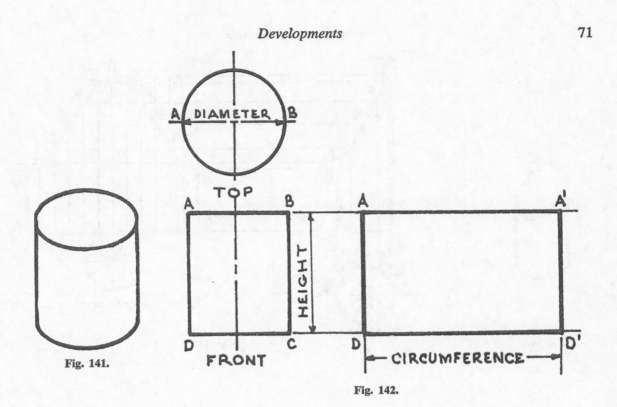

Fig. 141.

Fig. 142.

tical distance from point *C* to *B* in Fig. 142 represents the true height. The true width appears in the top view, Fig. 142, as the distance around the circle, called the **circumference.** By multiplying the distance from point *A* across to point *B*, or the **diameter,** by 22/7 or 3.1416, the circumference is obtained.

Drawing the Development. First draw the top and front views, as shown in Fig. 142. Extend horizontal lines from points *A* and *B*. At any convenient place, draw line *AD* parallel to line *BC*. Mark the length of the circumference along the horizontal line from point *A* to *A'*. Draw another vertical line from *A'* to *D'*. Complete the development with line *DD'*.

CYLINDER WITH SLANTED TOP

If the cylinder has a slanted top, as shown in Fig. 143, draw the front and bottom views as shown in Fig. 144.

Divide the circle in the bottom view into any number of equal parts, say twelve. Project lines from points *E*, *F*, *G*, *H*, and *J* in the bottom view up to the front view. These five projection lines will touch the slanted surface *AB* at five points. From these points and from points *A*, *B*, and *C* draw horizontal projection lines. Now draw the bottom or stretch-out line, as shown in Fig. 144. Divide this stretch-out line into the same number of equal parts that the circle was divided into, in this instance twelve. Vertical lines drawn from these twelve division points cross the horizontal projection lines. With a French curve connect these crossings to form a smooth curve.

TAPERING CYLINDER OR CONE

The development of a cone, illustrated in Fig. 145, is drawn in a manner very similar to the way a pyramid is drawn, except that the drawing of the cone presents fewer problems. The distance *AB* appears in its true length in the front view, as shown in Fig. 146. No matter which way the cone is revolved, the line *AB* appears in its true length. Take the length *AB* as a radius and draw an arc. See Fig. 146. Then draw a line from *A* to the arc, which gives us line *AB*. In order to place the circumference of

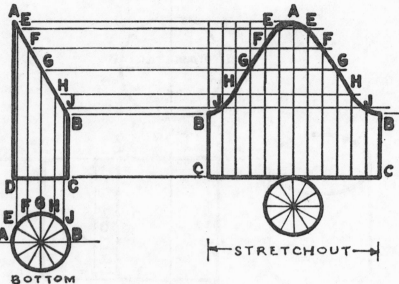

Fig. 143.

Fig. 144.

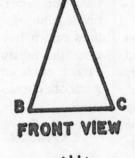

Fig. 145.

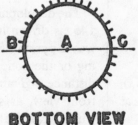

A

B   C

FRONT VIEW

B   A   C

BOTTOM VIEW

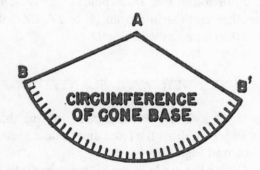

Fig. 146.

the base of the cone onto the arc, use the compass to divide the circumference into as many equal parts as possible. Then, starting from point *B* on the development, mark off the same number of equal parts on the arc as were marked off on the circumference in the bottom view. Connect *A* to the last mark on the arc to obtain *AB'*.

CONE WITH SLANTED TOP

To work out the development of the cone with the slanted top in Fig. 147, draw the top and the front views first. Start by drawing them as you would a complete cone. Then add the slanted top *DE* to the front view, and dot the lines from points *D* and *E* to the top of the cone at *A*, as shown in Fig. 148.

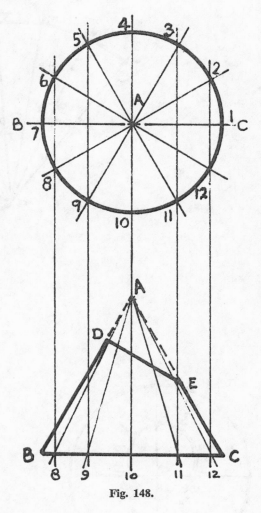

Fig. 148.

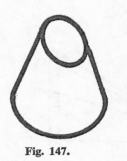

Fig. 147.

To show the shape of the slanted surface in the top view, divide the top view into any number of equal parts, say twelve. Where the division lines cross the circumference, drop projection lines to the base line, *BC*, in the front view, as in Fig. 148. These projection lines cross line *BC* at points *8, 9, 10, 11,* and *12*. From these points, draw lines to point *A* at the top of the cone. As these lines go to point *A*, they cross the slant line *DE* at points *F, G, H, J,* and *K*, as shown in Fig. 149. Draw lines straight up from these points and make them cross the division lines in the top view, as in Fig. 149. The crossings will create the points for the shape of the slanted top. Connect them with the French curve as in the top view of Fig. 150.

Drawing the Development. With line *AC* as a radius, draw the development of a complete cone, shown in Fig. 151. Note the points with the proper letters. Put point *C* on the outside. Line *CE* will be used for seaming because it is the shortest edge. Now the development line of the slanted top must be added. Since the slanted top appears as a curved line, its development is obtained by locating several points of the line. These points can only be found on lines in the front or top views that appear in their true lengths. Lines *AB* and *AC*, in the front view, show up in their true lengths. Therefore, the distance *AD* can be used to locate point *D*. On the development of the complete cone in Fig. 151, mark the distance *AD*, with the compass, on line *AB*.

Point *F'* in Fig. 151 does not appear in the front view because it is directly behind and at the same height as point *F*. The same holds true

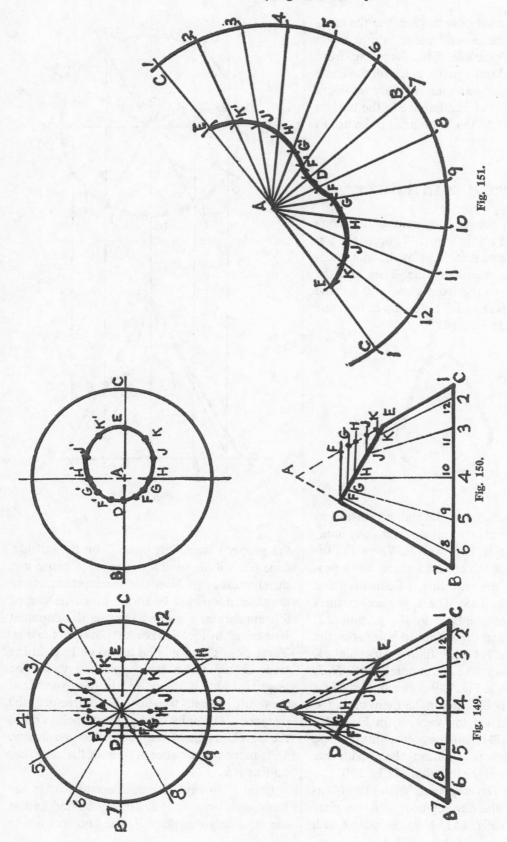

Fig. 151.

Fig. 150.

Fig. 149.

for the other pairs of points—*G and G',* *H and* *H', J and J', K and K'*—in Fig. 151. That means if you find the true location of one of the points in each pair you will have the true location of the other point of the pair.

Draw a horizontal line from point *F* in the front view and cross line *AC.* The distance *AF* on line *AC* is a true distance because line *AC* is a true-length line. Measure *AF* and mark point *F* on line *8A.* Mark point *F'* on line *6A* of the development. Draw similar horizontal lines from points *G, H, J,* and *K,* and locate them on their proper lines in the development, as shown in Fig. 151.

The distance *AE* on line *AC* is a true length. Locate it on both lines *1A* in the development. With the French curve, connect the points to complete the development.

CONE WITH UNEVENLY SLANTING SIDES

To draw a cone with unevenly slanting sides, shown in Fig. 152, first draw the top and front views, shown in Fig. 153. In the top view divide the circle, which is the bottom of the cone, into any number of equal parts. From point *A* draw lines to these divisions. Project lines from the point where these lines cross to the line *BC* in the front view. From the new points on line *BC* in the front view, draw lines to point *A.* At this point it is necessary to know the true lengths of these lines in order to draw the development.

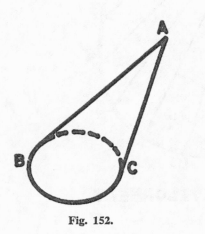

Fig. 152.

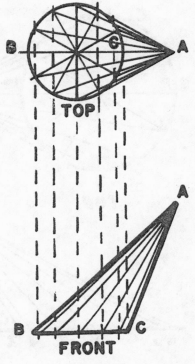

Fig. 153.

GETTING THE TRUE LENGTHS

The true lengths are obtained by first extending the line *BC* horizontally and then by dropping a perpendicular line to it from point *A,* as shown in Fig. 154. From point *A* in the top view, measure the distance to point *B* and mark it on the extended line *BC* from point *X* in the front view. Then measure the other distances from *A* in the top view to points *1, 2, 3, 4, 5,* and *6,* and mark these also from point *X* on the extended line *BC* in the front view. From these points on the extended line *BC,* draw lines to point *A.* The lines to point *A* are the true lengths needed to draw the development. See Fig. 154. When the true length of several lines is plotted on a drawing, it is called a true-length diagram.

Drawing the Development. To draw the development, **start with the shortest side, *A 12.*** Draw one line, *A 12,* **in its true length,** which may be obtained from Fig. 154. See Fig. 155.

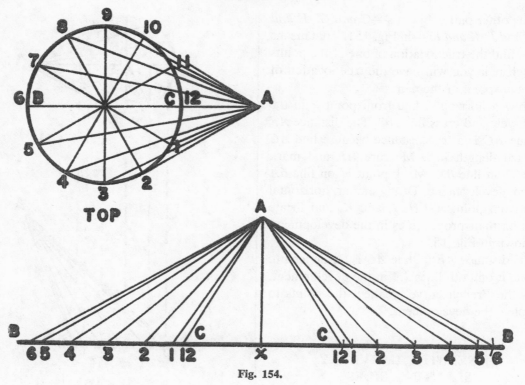

TOP

Fig. 154.

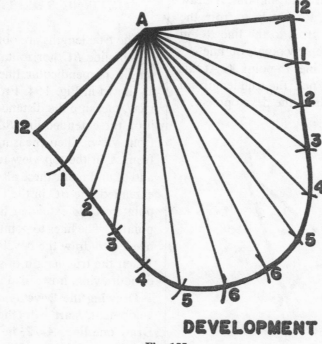

DEVELOPMENT

Fig. 155.

In the top view, Fig. 154, measure the length from point *12* to point *1* with the compass, and then draw an arc from point *12* in the development, Fig. 155. Now measure, in the front view, the distance of the true length *A1* with the compass, and then, from point *A* in the development, cross the arc just made from point *12*. This gives us the first point for the development, point *1*.

In the top view, Fig. 154, measure the distance from point *1* to point *2* with the compass, and mark an arc of that distance from point *1* in the development. Get the true length of line *A2* from the front view, Fig. 154. From point *A* in the development, Fig. 155, cross the arc made from point *1*. You now have point *2* for the development. Continue this process until you have located all the points up to line *A6*.

Both halves of the slanted cone are exactly the same size. When you have obtained line *A6*, merely start back again from line *A6* to get the other half of the development. After locating all the points, connect them with the French curve to complete the drawing.

OBJECTS WITH BOTH STRAIGHT AND CURVED SURFACES

Frequently an object must be drawn to connect a square or rectangular shape to a circular one, as in Fig. 156. This object can be made of one sheet of material. The development

can be drawn by combining the methods which we have already studied.

Start by drawing the top and front views, as shown in Fig. 157. *ABC* is flat, but *ABE* is curved. In the top view the lines *BG, CH, HG,* and *GB* appear in their true lengths. The circle also appears in its true size. However, the lines *AC, CD, DH, HF, FG, GE, FB,* and *BA* do not appear in their true dimensions in the top or front views. The true lengths may be obtained by first dividing the circle in the top view into any number of **equal** parts. Then project lines from points *10, 11, 12, 1,* and *2* to line *DAE* in the front view. From *DAE* in the front view,

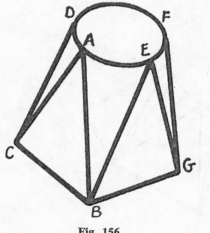

Fig. 156.

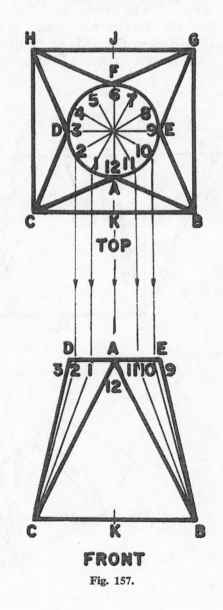

FRONT

Fig. 157.

draw lines from points *2, 1, 12, 11,* and *10* to points *B* and *C.* Now obtain the true lengths of these lines to points *B* and *C,* using the same method that you did when you worked with cones with unevenly slanted sides. Review Fig. 154.

To start the development, first draw the line *KC,* as in Fig. 159. Line *KC* appears in its true length in the top view, Fig. 157, as half the length of line *BC.* Now measure the true length of line *AC* in Fig. 158 with the compass, and mark it off from point *C* in Fig. 159. Complete the triangle by drawing line *AK.* Now get the true length of line *C1* in Fig. 158 and mark it off from point *C* in the development, Fig. 159.

Measure the distance from point *1* to point *12* in the top view, Fig. 157. In the development, cross the line *C1* with a small arc, using

point *12* as a center, to obtain the true location of point *1.* Connect points *1* and *12.* Measure the true length of line *C2* in Fig. 158 and mark it off from point *C* in the development. Measure the distance in the top view, Fig. 157, from point *1* to point *2* and cross line *C2* with an arc, using point *1* as a center. Connect lines *1* and *2* in Fig. 159. In Fig. 156 line *CD* has the same length as line *AC* in Fig. 159. Use line *AC* to obtain point *3.*

Now, from point *C* in the development, mark off the distance *CH* found in the top view. Obtain the true length of line *DH,* which is the same as the true length of line *AC.* From point *D* in the development, cross line *CH* to obtain the true location of point *H.* Continue this process all around the object to complete the development, as shown in Fig. 160.

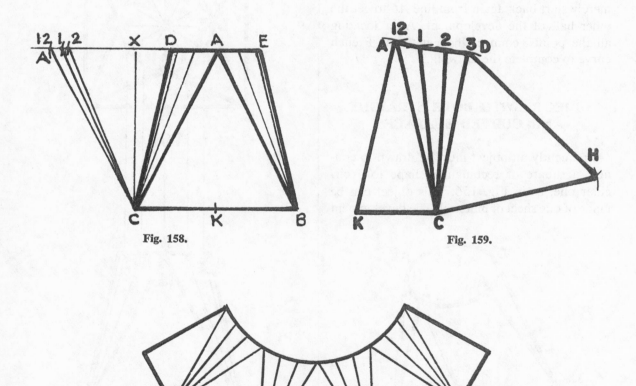

Fig. 158.

Fig. 159.

Fig. 160.

DEVELOPING A SPHERE

A sphere or ball curves in two general directions. It goes around from top to bottom, and around the middle at the same time, as shown in Fig. 161. When drawing developments, the sphere of the object is laid out flat. In order to draw the development of a sphere you have to flatten out the sphere in both directions. You must first take the surface from the very top, where the North Pole lies on the earth, for instance, and peel it off in a certain shape to the Equator. The procedure is repeated from the very bottom, or where the South Pole lies on the earth. See Fig. 162. The surface is still curving around the Equator. We must flatten the surface going around the Equator by peeling it from the center of the sphere, or the Equator, to obtain one piece containing all the north-south peeled shapes.

Drawing the Development. To make this development, draw a top and front view of the sphere, as shown in Fig. 163. You need only a semicircle in the top view, Fig. 163, because the sphere is symmetrical. In the top view, divide the semicircle in an equal number of parts. See points *A, B, C, D, E*. The shape *FAG* shows one of these equal parts.

In the front view, draw horizontal lines across the upper half. See points *1, 2, 3, 4, 5*. These lines represent circles going horizontally around the sphere. From the places where the horizontal lines touch the outside of the sphere, in the front view, project lines up to the line *GG* in the top view. These lines cross line *GG* at points *1* to *10*, as shown in Fig. 163.

In the top view, draw semicircles through the points *GG*, using point *F* as the center. The semicircles cross line *FA*. At these crossing points, *B, C, D, E*, project lines down to the front view. Where the projection lines cross the horizontal lines in the front view, draw a smooth curve.

Now draw a horizontal line to show the center of the sphere. See Fig. 164. Draw a line perpendicular to it. Measure the distance, with the compass, from point *1* to point *2* in the front view, and mark that on the development, using point *1* in the development as the center. Repeat this process for points *2* to *3, 3* to *4, 4* to *5*,

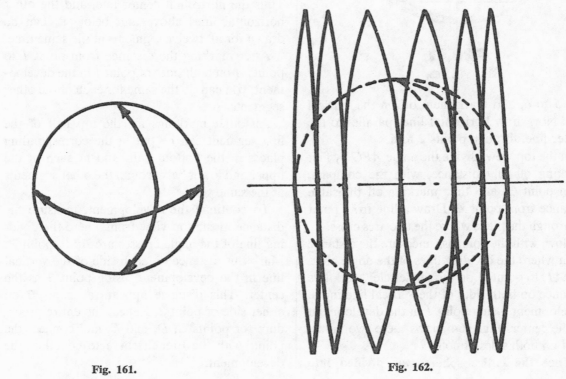

Fig. 161.   Fig. 162.

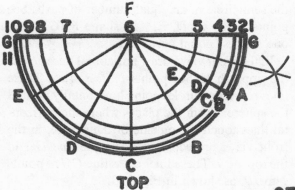

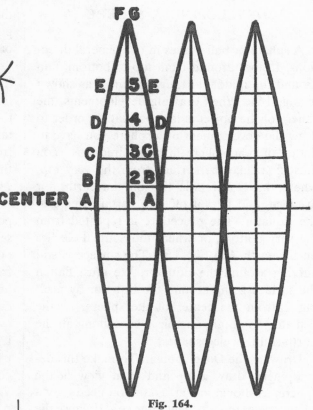

Fig. 164.

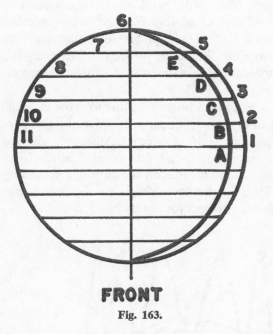

FRONT

Fig. 163.

and *5* to *6,* and mark them off on the vertical line. Now draw horizontal lines parallel to the center line, through points *2* to *6.*

In the top view, divide the angle *AFG* by first marking off any distance, with the compass, from point *G,* and then marking off the same distance from point *A.* Draw a line from point *F* through the place where the two arcs cross.

Now, with the compass, measure the distance from where the dividing lines of the angle cross arc *A1,* to point *A* on the semicircle. Mark this distance on both sides of the vertical line in the development, using point *1* in the development as the center. Fig. 164 shows these two points as *A* on both sides of point *1.*

Since the entire sphere was divided into

twelve equal parts, simply repeat each step you took in making the first segment twelve times. Thus the horizontal center line, and the other horizontal lines above and below it, can be drawn for all twelve segments at the same time.

After marking the distance from point *1* to points *A* on both sides of point *1* in the development, you can do the same for each of the other segments.

All other markings for the top half of the first segment can be put in the corresponding places in the bottom half, and in each of the upper and lower halves of all the other segments at the same time.

To continue the development, measure the distance in the top view from where the dividing line of the angle crosses arc *B2* to point *B.* Mark this distance on each side of the vertical line in the development, using point *2* as the center. This distance appears as point *B* on either side of point *2.* Repeat the entire procedure for points *C, D,* and *E,* and connect the points with the French curve to complete the development.

Practice Exercise No. 4

The following problems will test how well you understand the material covered in this chapter. Follow the directions given with each problem. After you have completed the required drawings check your solution with the answers shown in the answer section at the back of the book.

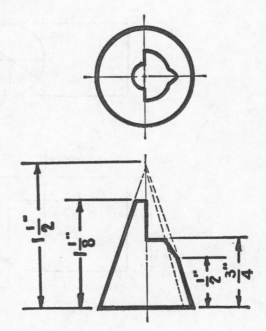

PROBLEM 2: Draw development.

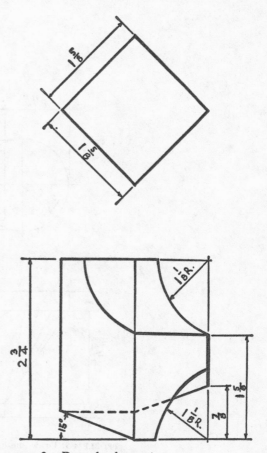

PROBLEM 3: Draw development.

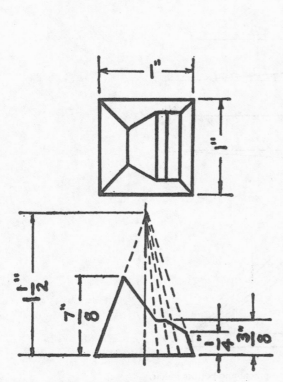

PROBLEM 1: Draw development.

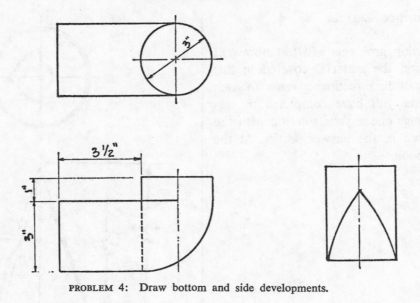

PROBLEM 4: Draw bottom and side developments.

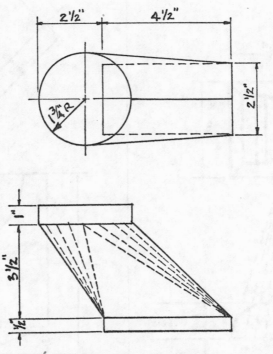

PROBLEM 5: Draw one-half development and bottom collar.

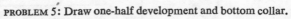

CHAPTER EIGHT

ISOMETRIC DRAWING

Many times a **three-dimensional** drawing will illustrate an object more clearly than an orthographic drawing. A three-dimensional drawing shows several sides of an object at the same time. There are many ways of drawing in this fashion. **Perspective drawing,** developed since the sixteenth century, still is most widely used for illustration. Mechanical drawings, however, are made for **construction,** not for illustration. True dimensions cannot be shown on perspective drawings and therefore a method of drawing known as **isometric drawing** is used.

BASIS OF ISOMETRIC DRAWING

In orthographic projection three views—the top, front, and side—are shown. Turn the object, shown in Fig. 165A, 45° in the front view, as shown in Fig. 165B, so that you see the corner line, *DE,* and the other two sides, *ADEH* and *CDEF,* instead of just one side. Then tilt the object up at the front corner, point *E,* so that the line *DE* makes an angle of 35°16′ with the vertical line *XY.* The edges *DC* and *AD* also make angles of 35°16′ with the horizontal line, as shown in Fig. 166. If you draw a front view of the object in this turned and tilted position, it will appear as shown in Fig. 167A.

The opposite lines on each side of the six sides of the object are parallel, as lines *AD* and *BC,* or lines *DE* and *CF.* This kind of turning and tilting is called **isometric projection,** as shown in Fig. 167A. The word **isometric** means the **same measurement.** By turning and tilting, the lines *AD, DC,* and *DE* became foreshortened to the same degree. An isometric projection can, therefore, be drawn to scale, since the foreshortened lines will appear in the same relation as they actually are in the object.

Isometric projection creates the odd angle of 35°16′. To make the drawing simpler, ignore

the angle of 35°16′ and substitute an angle of 30° instead. The change does not affect the drawing of true dimensions enough to make any difference. However, when the 30° angle is used, the drawing is called an **isometric drawing,** as shown in Fig. 167B, instead of an **isometric projection.**

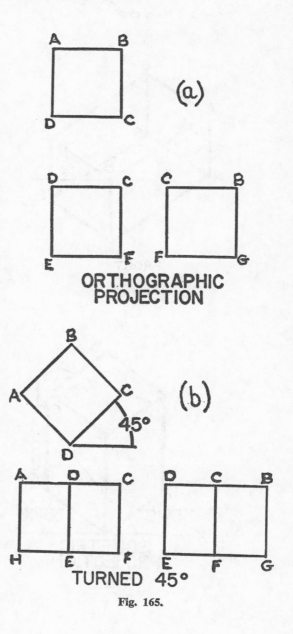

Fig. 165.

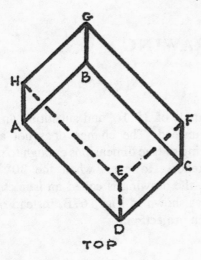

TOP

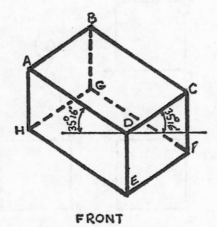

FRONT

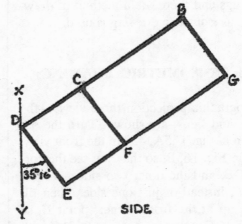

SIDE

Fig. 166.

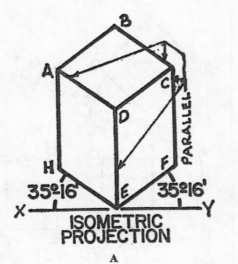

35°16'   35°16'

ISOMETRIC
PROJECTION

A

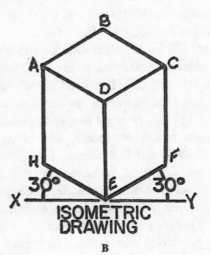

30°   30°

ISOMETRIC
DRAWING

B

Fig. 167.

ISOMETRIC DRAWING

Only one view appears in an isometric drawing. Start by drawing a horizontal line, *AB*, with the T square and then draw the vertical line, *CD*, perpendicular to it, as shown in Fig. 168. With the 30° triangle, draw lines *EF* and *GH*, making 30° angles with the horizontal line *AB*. All the lines cross each other at the same point, point *O*. Along line *EF* from point *O* mark off the length of the bottom edge of the object *O* to *K*. Mark off the length of the other bottom edge, *OL*, along line *GH*. Put the height, *OM*, of the corner of the object along line *CD*, from point *O*. At point *M* draw a line parallel to line *OF* with a 30° triangle, and another line parallel to line *OG* from point *M*.

Straight up from points *L* and *K*, draw vertical lines. These vertical lines cross the lines parallel to *OG* and *OF*, at points *N* and *P*. From *N* and *P* draw the rest of the top of the object. A line parallel to *CP* from *N* and a line from *P* parallel to *CN* will complete the top at *R*. Make the outline of the object, Fig. 168, darker and heavier than the other lines. Omit the hidden lines and you have an isometric drawing of the object.

NONISOMETRIC LINES

The outline of the object just drawn, Fig. 168, fell right on the isometric lines. Suppose some lines of the object go in another direction, as in Fig. 169A. To draw these nonisometric lines, *EF* and *HJ*, guide lines are needed. Start as you did in drawing the object in Fig. 168. All of the lines falling on the isometric lines can be located and drawn as before. Lines *EF* and *HJ* remain to be drawn. Simply extend the lines from points *D* and *G* and from points *A* and *K*. See Fig. 169B. Mark off the distance to *F* and *J* from points *A* and *K*, and then mark off the

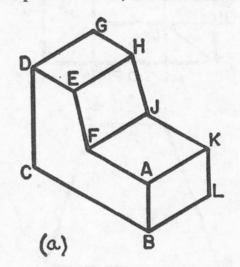

(a)

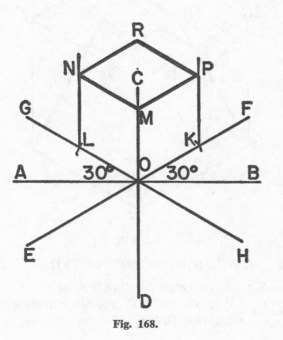

Fig. 168.

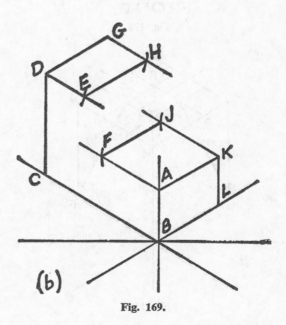

(b)

Fig. 169.

distance to *E* and *H* from points *D* and *G*. Connect the points to complete the isometric drawing shown in Fig. 169B.

The top and front views of a more complicated object appear in Fig. 170. To locate the points of the top surface, *ABCD*, draw an **isometric box** from which necessary measurements can be made. See Fig. 171. The height of

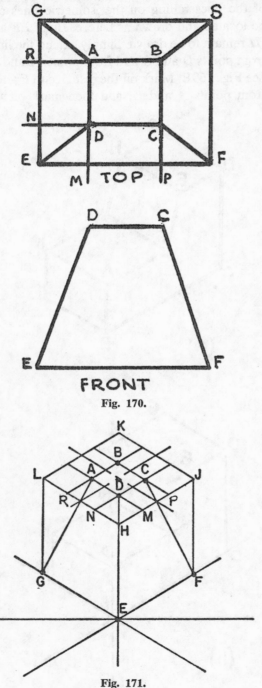

Fig. 170.

Fig. 171.

the box reaches the height of the object. Again start with the horizontal, vertical, and 30°-angle lines, with *E* as the corner point. Draw the isometric box lightly on the isometric lines. Extend line *AD* in the top view to cross line *EF* at *M*. See Fig. 170. Measure the distance *EM* and mark this distance from point *H* on line *HJ* of the isometric box, as point *M*, Fig. 171. Draw an isometric line from point *M* to line *LK*.

Extend line *BC* in the top view until it crosses line *EF* at *P*. Measure the distance *EP* and mark it on line *HJ* from point *H*. Draw an isometric line from point *P* across line *LK*.

In the top view, extend lines *DC* and *AB* across line *EG*. From point *E* measure the distances *EN* and *ER* and transfer them to the isometric box on line *HL* from point *H*. Draw isometric lines from points *N* and *R* across line *JK*.

The four points *A*, *B*, *C*, and *D* will appear where the lines intersect. See Fig. 171. Draw lines to the four points and to points *G* and *F* to complete the isometric drawing, as shown in Fig. 172.

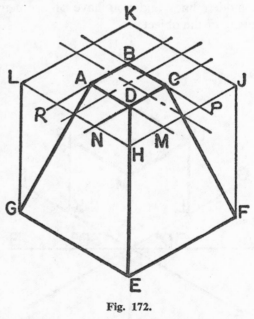

Fig. 172.

ANGLES IN ISOMETRIC VIEWS

Since in isometric drawing the size of angles is limited to 30° and 60°, any other angle will appear distorted. In Fig. 167B, for instance, the

90° angle of the box, angle *ADC*, appears as a 120° angle on the isometric drawing. Angles on the isometric view do not reflect their true sizes. Fig. 173A shows the method that is used when working with an object having an angle in its shape.

Draw the isometric box as previously explained in our discussion of Fig. 171. Show an orthographic view of the object, as in Fig. 173A. Draw a horizontal line running through *B* and parallel to *ED* in the isometric box, and going through point *G*, as shown in Fig. 173B. Draw a vertical line up from *B* in the orthographic view, Fig. 173A, and cross it with a line from *A to C*. This gives us point *H*. In the isometric view, Fig. 173B, mark off the distance *A to H* and drop a vertical line crossing line *GK*. This gives us point *B*. Connect lines to

points *A and B* and *B and C*, which gives us the isometric drawing of the angle at *B,* as shown in Fig. 173C.

CIRCLES AND CURVES IN ISOMETRIC DRAWINGS

Curves. Guide lines are used to show a curve in an isometric drawing. The guide lines help locate the points of the curve. First draw an orthographic front view, as shown in Fig. 174A. Then add any number of guide lines, such as *DD′* and *EE′*, etc., from one edge of the object to the curved edge. The straight edge will fall on an isometric line, from which measurements can be taken. Now draw the vertical, horizontal, and 30° lines, as shown in Fig. 174B. Measure the distance *AB* in the front view, Fig. 174A, and mark it off on the isometric draw-

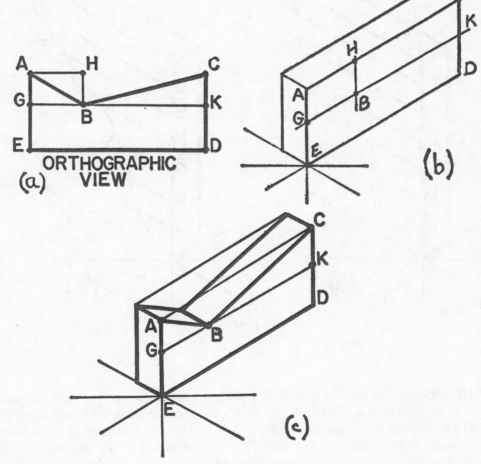

Fig. 173.

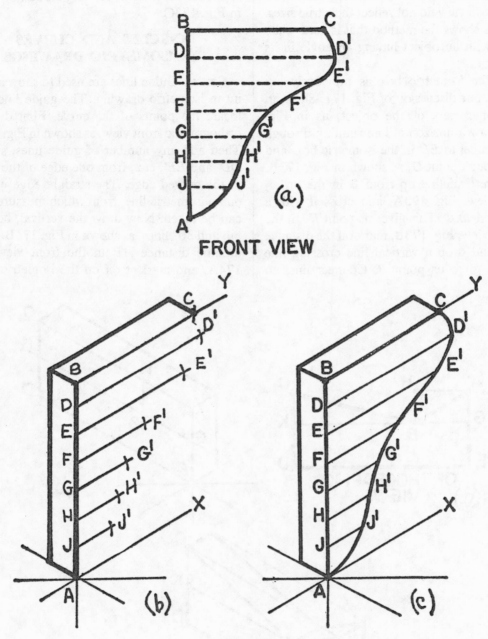

FRONT VIEW

Fig. 174.

ing, Fig. 174B. Measure the vertical distances from points *B to D, D to E, E to F*, etc., on the front view. Mark them on the isometric drawing. Draw isometric lines parallel to line *AX* from the points mentioned above. Now measure the distances from points *B to C, D to D', E to E'*, etc., on the front view and mark them on the corresponding lines of the isometric drawing. Connect the points with the French curve to obtain the isometric drawing. See Fig. 174C.

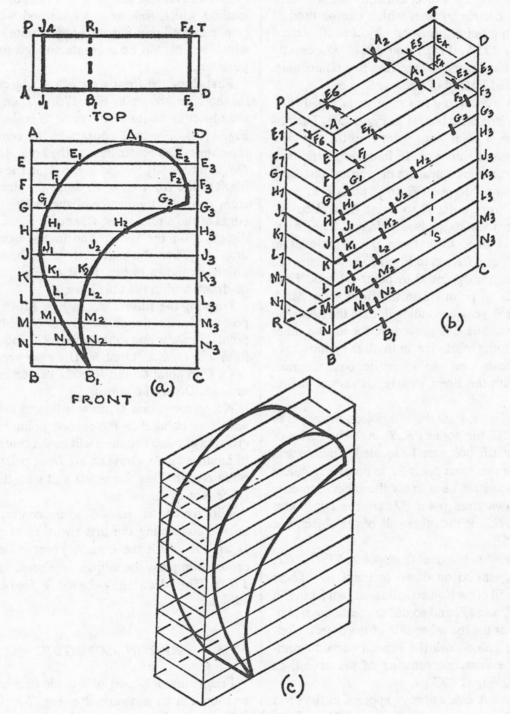

Fig. 175.

If the original object contains neither vertical nor horizontal lines which can be used as measuring points, draw an object as illustrated in Figs. 175A, 175B, and 175C. Do exactly as you did before, except that two points must be located instead of one.

Draw a front and top view, as in Fig. 175A, then an isometric box, as in Fig. 175B. The distance from point *A* to point *P* in the top view is the same distance as *AP* in the isometric box. In the front view, measure the vertical distances from point *A* to point *E,* from point *E* to point *F,* and from point *F* to point *G,* and so on. Transfer them to the isometric box. Then draw isometric lines from these points parallel to lines *BC* and *BR,* as in Fig. 175B. Measure the distance from point *E* to point *E1,* and from point *E1* to point *E2* in the front view. Mark them on the corresponding lines in the isometric box. Continue this procedure for all the other points in the front view until all the points have been located on the isometric box. Connect them with the French curve, as shown in Fig. 175C.

In order to draw the full object, as shown in Fig. 175C, the points *PRST* on the back side of the isometric box must be located. An isometric line is drawn from point *E1* to point *E2.* Where these isometric lines cross the other isometric line drawn from point *E7,* are the two points *E6* and *E5* for the other side of the object. See Fig. 175B.

Draw isometric lines from points *F1, G1, H1,* and *J1,* and so on down to point *B1.* These points will cross isometric lines drawn from *F7, G7, H7,* and *J7,* and so on, to create the rest of the points for the other side of the object. Connect the points with the French curve to complete the isometric drawing of the object, as shown in Fig. 175C.

Circles. A circle always appears in the form of an ellipse in isometric drawings. The ellipses conform to the isometric angles. A template, which contains many sizes of ellipses, may be used to trace an ellipse of the desired size. Templates may be purchased at most art or drafting stores. If the template does not contain the proper size ellipse, you can draw one, using the method which follows. This method will not give a true ellipse, but the approximate ellipse which results will be accurate enough for our purposes.

First draw the front view of the object with the hole, as shown in Fig. 176A. Then draw the object in an isometric view, to scale, as in Fig. 176B. The hole appears in the center of the object. The center lines of the hole, *AC* and *DB,* can be easily located in the isometric drawing. On the front view, measure the distances from point *O,* the center of the circle, to the edges of the circle, *A, B, C,* and *D.* Mark these distances on the center line in the isometric drawing. Then draw lines through these four points, as shown in Fig. 176B. The ellipse will be drawn within the four lines.

Drawing the Ellipse. As in Fig. 176C, from point *D* draw a line to the corner *G,* and from point *E* draw a line to the corner *C.* These two lines will cross at point *R.* Draw an arc from point *D* to point *C* with *R* as the center and the distance *DR* as the radius.

Now, from point *G* draw a line to point *A* and from point *B* to the corner, point *E.* The crossing of these two lines will be the center, *R',* of another circle. Draw an arc from point *A* to point *B* with *R'* as the center and with the distance *R'A* as the radius.

With the corner, point *G,* as the center, draw an arc connecting the first two arcs at points *D* and *A.* With the corner, point *E,* as the center, complete the ellipse, as shown in Fig. 176D. This drawing is called a **four-center ellipse.**

ARC IN ISOMETRIC

Frequently only part of a circle or arc must be shown in an isometric drawing. When such is the case only part of an ellipse must be constructed to show the necessary part of the circle or the arc.

Follow the same procedure as you used in drawing an ellipse. Fig. 177A shows the top and front views of an object with an arc. The

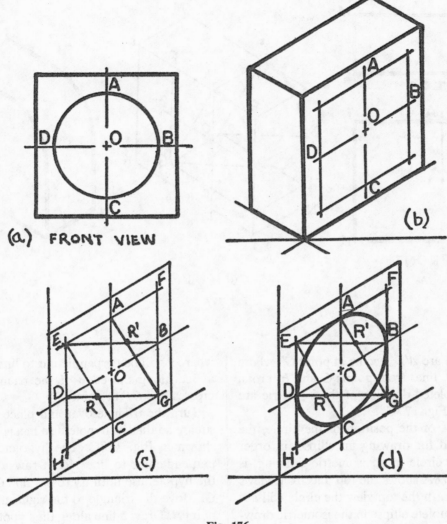

(a) FRONT VIEW

(b)

(c)

(d)

Fig. 176.

arc, which appears here as a *quarter of a circle*, will become a *quarter of an ellipse* in the isometric drawing. The length of the radius, *OA*, and the location of the center of the arc, point *O*, must be known. Draw an isometric box for the object, as shown in Fig. 177B. Locate the center of the arc, point *O*, by measuring line *JD* on the front view. Mark it on the vertical line, *JE*, in the isometric box. Then draw an isometric line from point *D*. Measure the distance from point *E*, in the front view, to point *A*, and mark it on the isometric line, *EL*. Drop a vertical line from point *A* on the isometric box.

Point *O* occurs where the two lines cross. The length of the vertical line *AC* will be twice as long as line *AO*, and line *DB* will be twice as long as line *DO* because the shorter lines are radii and the longer lines are diameters. This is shown in Fig. 177B.

Working with Fig. 177B, draw lines from point *D* to point *G* and from point *A* to point *G*. With *G* as a center, draw an arc from points *A* to *D* to obtain the isometric arc.

Construct the arc *A'D'* by drawing isometric lines from points *G* and *H*. From point *H'* draw another isometric line to the right. The

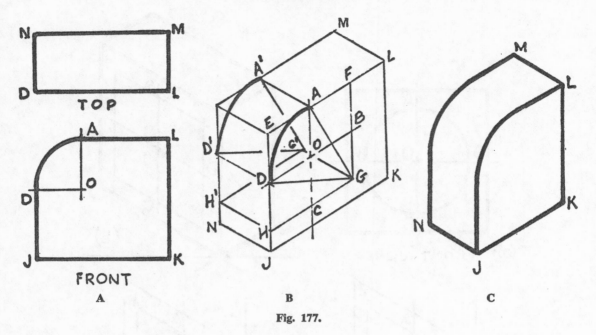

Fig. 177.

center of the arc *A'D'* occurs at point *G'*, where the isometric lines from points *G* and *H'* cross.

The completed drawing of the isometric arc is shown in Fig. 177C.

Regardless of the position of the circle, the same method for drawing an ellipse, in order to show the circle in an isometric drawing, is used. Fig. 178A shows the top and front views of a cylinder. In the top view the circle will appear as a complete ellipse in the isometric drawing, while the bottom of the cylinder will appear as half an ellipse. Short cuts can be used to make both drawings.

Drawing the Ellipse. Draw the three basic isometric lines or axes as lines *AC*, *BD*, and *EF* in Fig. 178B. Lines *AC* and *BD* will equal the length of the diameter of the circle, as shown in the top view. Line *OF* in the isometric drawing will be the same length as line *OF* in the front view.

Place the 30° triangle at point *C*, as shown in Fig. 178B. Line *DB* will be perpendicular to the long edge or **hypotenuse** of the triangle. Draw a line along the hypotenuse until it crosses line *OF* at point *E*.

Move the triangle along the T square, to the right, until the hypotenuse meets point *A*,

where it will be perpendicular to line *DB*. Draw a line along the edge of the triangle until it crosses line *OF* at point *G*.

Turn the triangle over and place it on the T square so that its hypotenuse meets point *D*, as shown in Fig. 178C. The hypotenuse will be perpendicular to line *AC*. Draw a line along the hypotenuse until it crosses line *OF* at point *G*. Move the triangle to the right until it meets point *B*. Draw a line along the hypotenuse until it meets point *E*.

In Fig. 178D lines *EC* and *DG* cross at point *R*, and lines *EB* and *AG* cross at point *R1*. From points *R* and *R1* drop vertical lines. Draw a horizontal line, *HJ*, at point *F*, to cross these vertical lines at points *R2* and *R3*. With distance *RC* as a *radius*, draw an arc from the center, *R*, to points *C* and *D*, and with *R1* as the center, draw an arc to points *A* and *B*.

From points *B* and *C*, drop vertical lines. Again, with distance *RC* as a radius, and point *R2* as the center, draw an arc from the horizontal line *HJ* to the vertical line coming from point *B*. With point *R3* as a center, draw a similar arc from line *HJ* to the vertical line coming from point *C*.

Put the hypotenuse of the triangle at point

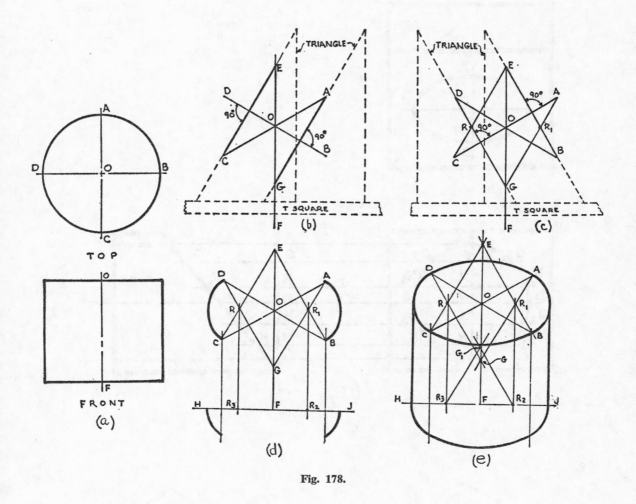

Fig. 178.

R2 and draw a line to cross line *OF* at point *G1*, as shown in Fig. 178E. With point *G1* as a center and line *EB* as a radius, complete the lower half-ellipse. Use the same radius, *EB,* to complete the upper ellipse by drawing arcs from points *E* and *G.*

Draw vertical lines connecting the upper and lower ellipses, to complete the isometric drawing of the cylinder shown in Fig. 178E.

CIRCLES ON NONISOMETRIC LINES

A circle in an object may not fall on an isometric line, as shown in Fig. 179A. To draw such an object isometrically, the points of the circle must be located on an isometric box and

be transferred from there to their proper positions in the isometric drawing.

DRAWING THE NONISOMETRIC CIRCLE

Draw an isometric box based on the dimensions of the object. See Fig. 179B. Add the center lines of the circle to the box and locate the edges of the circle, points *A, B, C,* and *D.*

In the side view, Fig. 179A, draw a vertical line, *EF,* which represents the *edge of the front side of the isometric box.* In the front view, Fig. 179A, draw horizontal lines at any desired location, across the circle, and extend them to the side view.

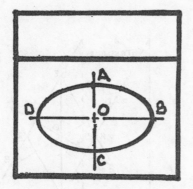

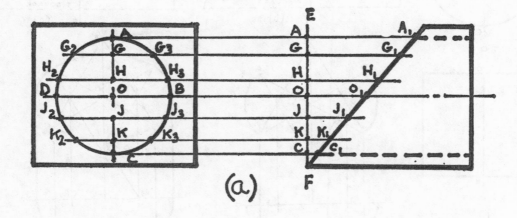

(a)

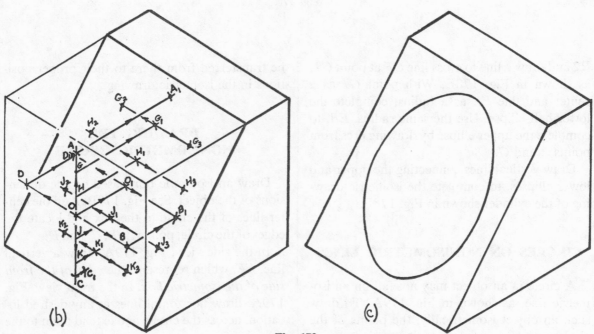

(b)   (c)

Fig. 179.

From the center of the circle, point *O*, in the front view, measure the distances up and down line *AC*, to points *G, H, J*, and *K*. Transfer each of these distances to the line *AC* on the isometric box and draw isometric lines from each point, as shown in Fig. 179B.

In the side view, Fig. 179A, measure each of the distances from line *EF* to the slanting line as *A* to *A1, G* to *G1*, etc., and mark them on the corresponding isometric line of the isometric box. This will give you points *G1, H1, O1, J1, K1*, and *C1*, as shown in Fig. 179B.

In the front view, Fig. 179A, measure the horizontal distance from points *G* to *G3*. From point *G1* on the isometric box, Fig. 179B, mark this distance on both sides of point *G1*. You now have points *G2* and *G3*. Do the same with the other horizontal distances in the front view to obtain points *H2, H3, B1, D1, J2, J3, K2*, and *K3*. Connect the points with the French curve and you have the completed drawing as shown in Fig. 179C.

VARIOUS POSITIONS OF ISOMETRIC AXES

Up to now we have been working with drawings of isometric axes as shown in Fig. 180A. Actually, the axes can be drawn in several positions since an object may reveal its details if turned at some isometric angle other than the one we have been working with up to this point. Figs. 180B, 180C, and 180D show other posi-

tions of the isometric axes. Naturally, the position to be used is determined by a decision as to which position will show either the whole object or some particular detail most clearly.

The principles of isometric construction just described will hold true for any of the positions in which you place the isometric axes.

The reversed position, Fig. 180B, has the 30° angle below, instead of above, the horizontal line. The reversed position is used a great deal in architectural detail drawings. The object appears as if you were looking up at it, thus revealing many parts not seen from any other view.

ISOMETRIC SECTIONS

Full Sections. It may be necessary to show **sections** of an object as an isometric drawing. In Chapter Two we discussed the methods of drawing **orthographic sections.** In drawing an **isometric section,** first draw an orthographic section, as shown in Fig. 181A. Then draw the isometric view of the section face only, as shown in Fig. 181B. To complete the isometric drawing, add the rest of the section to the drawing. See Fig. 181C. Crosshatch the cut surfaces at a 60° angle.

Half Sections. When drawing the isometric view of a half section, make the complete object in isometric view, then remove the front quarter. Fig. 182A shows the front view of an

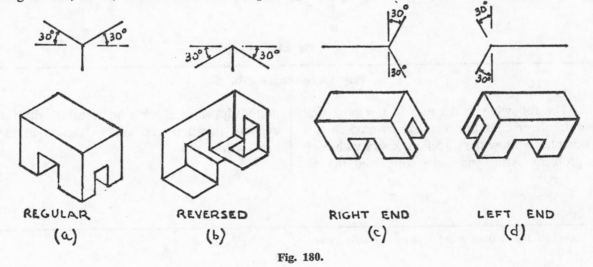

REGULAR   REVERSED   RIGHT END   LEFT END

(a)   (b)   (c)   (d)

Fig. 180.

object, and Fig. 182B shows the isometric view of the complete object. Isometric center lines can be drawn on the top of the complete isometric view. By cutting the complete isometric view along the center lines, *OA* and *OB*, you will have an isometric half section, as shown in Fig. 182C. Crosshatch the cut surfaces at a 60° angle.

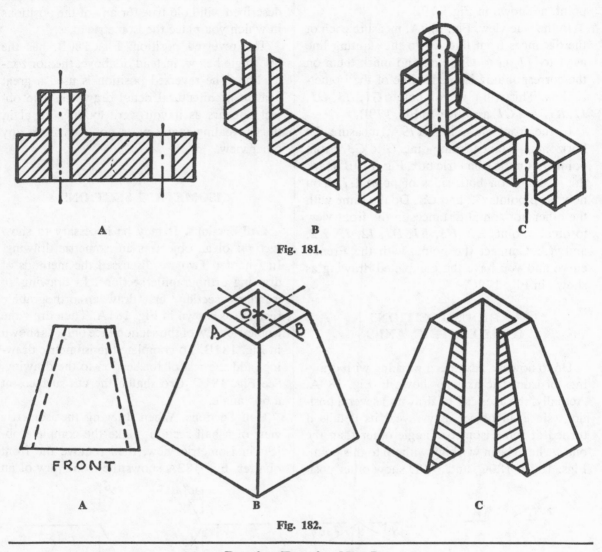

A

B

Fig. 181.

C

A

FRONT

B

Fig. 182.

C

Practice Exercise No. 5

The following problems will test how well you understand the material covered in this chapter. Follow the directions given with each problem. After you have completed the required drawings check your solution with the answers shown in the answer section at the back of the book.

PROBLEMS 1 & 2 (next page):   Draw isometric view.

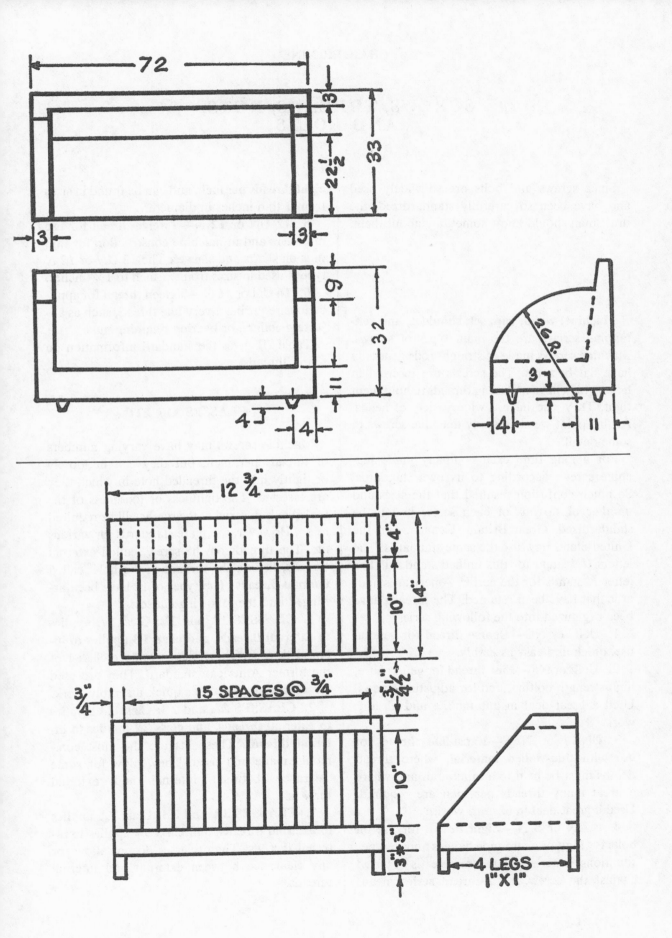

CHAPTER NINE

SCREWS, THREADS, NUTS, AND BOLTS

Since screws and bolts are so widely used and have been so carefully standardized, the draftsman should know something about them.

SCREWS

Machine, wood, cap, set, shoulder, and **self-tapping** screws are the main types of screws. Machine screws made of straight rods generally have flat bottoms. The screws are inserted in holes which have matching threads to hold them tight. They also have a wide variety of heads. At this point we shall study machine screws in some detail.

For a long time each company made machine screws according to its own standards. So much confusion resulted that the **sizes** and **number of threads** of each screw had to be standardized. Great Britain, Canada, and the United States now use the same standards. The letters *UN* stand for this unified standard. The letter *N* stands for the earlier American standards that have been retained. The threads have been organized into the following series:

1. **UNC** or **NC—Coarse thread** for general use, **quick and easy assembly.**

2. **UNF** or **NF—Fine thread** for general use, high-strength bolting, and for adjusting screws. Used a great deal in **automobile and aircraft** work.

3. **UNEF** or **NEF—Extra-fine thread** for use with **thin-walled material,** where thread depths must be held to a minimum, and where a great many threads per inch are required. Used a great deal in **aircraft** work.

4. **8 UN** or **8 N—Eight thread** for use on bolts for high pressure pipe flanges and in other situations where great pressure is exerted against the screw. **Eight-thread series** means **eight threads per inch,** and can be found in sizes from 1 to 6 inches in diameter.

5. **12 UN** or **12 N—Twelve thread** for use in **boilers** and in **machine construction** for **thin nuts** on **shafts** and **sleeves.** The *12 UN* or *12 N* type is used in sizes from ½ inch to 1¾ inches.

6. **16 UN** or **16 N—Sixteen thread** for applications requiring a very **fine** thread, such as **adjusting collars,** or **bearing retaining nuts.**

Table III gives the standard information on screw threads.

CLASSES OF FIT

Machine screws may have varying numbers of threads per inch, but they can fit loosely or tightly into the threaded hole in which they are inserted. The tightness or looseness of the fit depends upon what the screw will be required to do. However, there are standards for various fits. The threads on the screw, called **external threads,** fit into the threads in the hole, called **internal threads.** The types of fit have been arranged into the following classes:

1. **CLASSES 1A and 1B: Class 1A applies to external threads, and class 1B applies to internal threads.** These classes replace class 1 of the former American standard. They are used for **ordinance** and other quick assembly work.

2. **CLASSES 2A and 2B: Class 2A applies to external threads, and class 2B applies to internal threads.** These fits are the most commonly made and used. They allow for some clearance between internal and external threads.

3. **CLASSES 3A and 3B: Class 3A applies to external threads, and class 3B applies to internal threads.** These classes do not allow for any clearance between external and internal threads.

SCREW THREAD SERIES

Size	Basic Major Dia	Threads Per Inch						Size
		Coarse (UNC or NC)	Fine (UNF or NF)	Extra Fine (UNEF or NEF)	8 Thread Series (UN, N or NS)	12 Thread Series (UN or N)	16 Thread Series (UN or N)	
0	0.0600	—	80	—	—	—	—	0
1	0.0730	64	72	—	—	—	—	1
2	0.0860	56	64	—	—	—	—	2
3	0.0990	48	56	—	—	—	—	3
4	0.1120	40	48	—	—	—	—	4
5	0.1250	40	44	—	—	—	—	5
6	0.1380	32	40	—	—	—	—	6
8	0.1640	32	36	—	—	—	—	8
10	0.1900	24	32	—	—	—	—	10
12	0.2160	24	28	32	—	—	—	12
1/4	0.2500	20	28	32	—	—	—	1/4
5/16	0.3125	18	24	32	—	—	—	5/16
3/8	0.3750	16	24	32	—	—	—	3/8
7/16	0.4375	14	20	28	—	—	—	7/16
1/2	0.5000	13	20	28	—	12	—	1/2
9/16	0.5625	12	18	24	—	12	—	9/16
5/8	0.6250	11	18	24	—	12	—	5/8
11/16	0.6875	—	—	24	—	12	—	11/16
3/4	0.7500	10	16	20	—	12	16	3/4
13/16	0.8125	—	—	20	—	12	16	13/16
7/8	0.8750	9	14	20	—	12	16	7/8
15/16	0.9375	—	—	20	—	12	16	15/16
1	1.0000	—	14	—	—	—	—	1
1	1.0000	8	12	20	8	12	16	1
1 1/16	1.0625	—	—	18	—	12	16	1 1/16
1 1/8	1.1250	7	12	18	8	12	16	1 1/8
1 3/16	1.1875	—	—	18	—	12	16	1 3/16
1 1/4	1.2500	7	12	18	8	12	16	1 1/4
1 5/16	1.3125	—	—	18	—	12	16	1 5/16
1 3/8	1.3750	6	12	18	8	12	16	1 3/8
1 7/16	1.4375	—	—	18	—	12	16	1 7/16
1 1/2	1.5000	6	12	18	8	12	16	1 1/2
1 9/16	1.5625	—	—	18	—	—	16	1 9/16
1 5/8	1.6250	—	—	18	8	12	16	1 5/8
1 11/16	1.6875	—	—	18	—	—	16	1 11/16
1 3/4	1.7500	5	—	16	8	12	16	1 3/4
1 13/16	1.8125	—	—	—	—	—	16	1 13/16
1 7/8	1.8750	—	—	—	8	12	16	1 7/8
1 15/16	1.9375	—	—	—	—	—	16	1 15/16
2	2.0000	4 1/2	—	16	8	12	16	2
2 1/16	2.0625	—	—	—	8	—	16	2 1/16
2 1/8	2.1250	—	—	—	8	12	16	2 1/8
2 3/16	2.1875	—	—	—	—	—	16	2 3/16
2 1/4	2.2500	4 1/2	—	—	8	12	16	2 1/4
2 5/16	2.3125	—	—	—	—	—	16	2 5/16
2 3/8	2.3750	—	—	—	—	12	16	2 3/8
2 7/16	2.4375	—	—	—	—	—	16	2 7/16
2 1/2	2.5000	4	—	—	8	12	16	2 1/2
2 5/8	2.6250	—	—	—	8	12	16	2 5/8
2 3/4	2.7500	4	—	—	8	12	16	2 3/4
2 7/8	2.8750	—	—	—	—	12	16	2 7/8
3	3.0000	4	—	—	8	12	16	3
3 1/8	3.1250	—	—	—	8	12	16	3 1/8
3 1/4	3.2500	4	—	—	8	12	16	3 1/4
3 3/8	3.3750	—	—	—	—	12	16	3 3/8
3 1/2	3.5000	4	—	—	8	12	16	3 1/2
3 5/8	3.6250	—	—	—	8	12	16	3 5/8
3 3/4	3.7500	4	—	—	8	12	16	3 3/4
3 7/8	3.8750	—	—	—	—	12	16	3 7/8
4	4.0000	4	—	—	8	12	16	4
4 1/4	4.2500	—	—	—	8	12	16	4 1/4
4 1/2	4.5000	—	—	—	8	12	16	4 1/2
4 3/4	4.7500	—	—	—	8	12	16	4 3/4
5	5.0000	—	—	—	8	12	16	5
5 1/4	5.2500	—	—	—	8	12	16	5 1/4
5 1/2	5.5000	—	—	—	8	12	16	5 1/2
5 3/4	5.7500	—	—	—	8	12	16	5 3/4
6	6.0000	—	—	—	8	12	16	6

Table III.

4. **CLASSES 2 and 3:** These American standards have been retained. These classes do not allow for any clearance between external and internal threads.

DESIGNATING A SCREW THREAD

The draftsman notes the type of thread to be used by giving the nominal size or major diameter of the threaded part, the number of threads per inch, the thread series, and the class of screw-thread fit. For example, the draftsman might need a ¼"-20 UNC-2A. The proper method of designating a screw thread is shown in Figs. 183A and 183B.

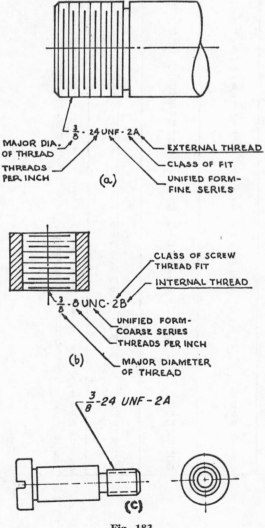

(a)

(b)

(c)

Fig. 183.

The draftsman often decides what screws should be used. He determines what size or diameter of screw will be suitable, and he looks in the table of screws to find the particular size that will be required, for example, ⅜". He decides to use a fine thread and looks under the fine-thread listings. The table of screws shows that a fine thread with a ⅜" diameter has 24 threads per inch. The draftsman thinks a class 2A fit will suffice and therefore indicates on his drawing that a particular screw must be ⅜"-24 UNF-2A. This designation is shown in Fig. 183C.

If the threads are to be **left-handed** (ordinary screws go into a hole when turned clockwise), the letters *LH* must be added to the end of the designation.

VARIATIONS IN DESIGNATING SCREWS

The designation of a screw may vary. One way of designating a screw is shown in the above example. See Fig. 183C. Here are other ways:

> 10-32 NF-2PD .1697-.1670
> ¼-20 UNC-2A
> 2¼-8 N-3 PD 2.1688-2.1611
> 1⅝-IONS-2 PD 1.5600-1.5532
> ¼-20 UNC-2A LH PD .2164-.2127
> ¼-20 UNC-2A TRIPLE

In the above examples, the letters *PD* stand for PITCH DIAMETER. **The pitch of a thread is the distance from one point on a thread to a corresponding point on the next thread,** shown as p in Fig. 184. The **pitch** equals one inch divided by the number of threads per inch. **The pitch diameter is the diameter of the pitch circle. The pitch circle is an imaginary circle whose circumference passes through a point where the thickness of thread equals the distance between threads.** In Fig. 185, the pitch circle would be located at the point where *AB* equals *BC*. The pitch diameter may be expressed in fractions or decimals. The numbers which follow the letters *PD* in the above examples indicate how small and how large the pitch diameters may be.

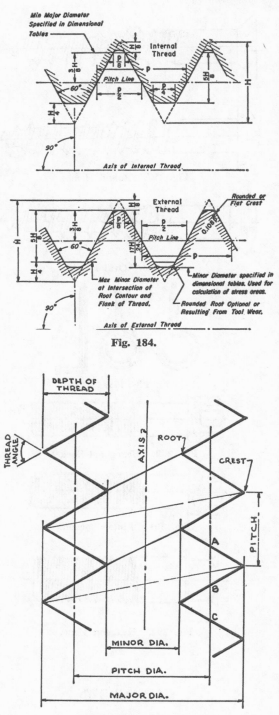

Fig. 184.

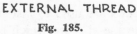

EXTERNAL THREAD

Fig. 185.

In the final example above, the word TRIPLE refers to the **lead. The lead of a screw is the distance the screw travels when it is given one complete turn.** When the screw moves only **one**

thread, the lead equals the pitch, and the thread is said to be **single lead.** When the screw moves **two threads,** the lead is **twice** the pitch and is said to be **double lead.** The word TRIPLE, in the designation above, means the thread should be **three times the pitch.**

Coated-Thread Designations. If the screw is to be **coated** or **plated** with some material, the screw is designated as, for example:

> ¼-20 UNC-2A
> PD 0.2164-0.2127
> BEFORE COATING
> PD 0.2175 MAX.
> AFTER COATING

If the screw is to be coated and no allowance is made for the pitch diameter, the screw is designated as, for example:

> ¼-28 UNF-3A
> PD 0.2268-0.2243
> AFTER COATING

Designations for Diameter Variations. If the major diameter variation of an external thread is to vary from the standard, the screw is designated as, for example:

> ⅜-24 UNF-3A MOD.
> MAJOR DIAMETER 0.3720-0.3648 MOD.

If the minor diameter of an internal thread is to vary from the standard, the screw is designated as, for example:

> ⅜-24 UNF-2B MOD.
> MINOR DIAMETER 0.330-0.336 MOD.

Special Thread Designations. If a special **thread form** is required, the screw is designated as, for example:

> 7/16-24 AM. NAT. FORM-SPECIAL
> MAJOR DIAMETER 0.4340-0.4280 SPL
> PD 0.4065-0.4025 SPL
> GO GAGE LENGTH ⅜-inch MIN.

If a special **thread angle** is required, the screw is designated as, for example:

> ⅞-18 SPECIAL FORM
> THREAD ANGLE 60°
> MAJOR DIAMETER 0.8750-0.8668
> PITCH DIAMETER 0.8384-0.8343
> MAX. MINOR DIA. 0.8068 (as gaged)
> GO GAGE LENGTH 11/16-inch MIN.

MACHINE SCREW NAMES AND HEADS

Names of machine screws depend on the type of machine-screw head. Figs. 186 through 195 illustrate the various types of machine-screw heads.

WOOD SCREWS

Wood screws have the same nominal sizes as machine screws, except that wood screws come in a great variety of threads per inch. Wood screws are denoted by number only, ranging from 0 to 24. The higher the number the fewer the threads per inch. Figs. 196 to 198 contain the data for standard wood screws. Wood screws are available in steel, bronze, or brass, with nickel, chrome, or other finishes.

Fig. 199 illustrates other types of screw fasteners used in wood and other soft materials.

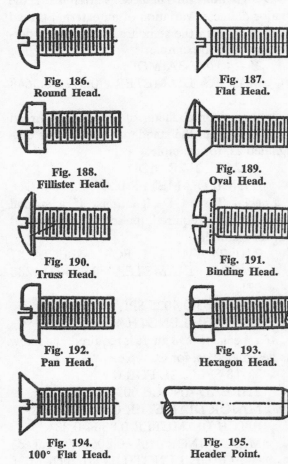

**Fig. 186.
Round Head.**

**Fig. 187.
Flat Head.**

**Fig. 188.
Fillister Head.**

**Fig. 189.
Oval Head.**

**Fig. 190.
Truss Head.**

**Fig. 191.
Binding Head.**

**Fig. 192.
Pan Head.**

**Fig. 193.
Hexagon Head.**

**Fig. 194.
100° Flat Head.**

**Fig. 195.
Header Point.**

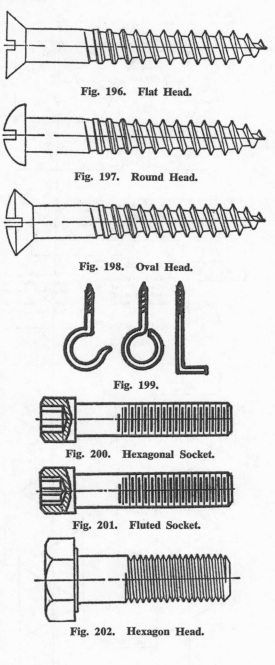

Fig. 196.   Flat Head.

Fig. 197.   Round Head.

Fig. 198.   Oval Head.

Fig. 199.

Fig. 200.   Hexagonal Socket.

Fig. 201.   Fluted Socket.

Fig. 202.   Hexagon Head.

CAP SCREWS

Cap screws are used to fasten two pieces of metal. The metal piece which is closest to the screw has no internal threads. Figs. 200 through 205 illustrate the various standard cap screws. Very often cap screws are used on machine tools requiring close dimensions.

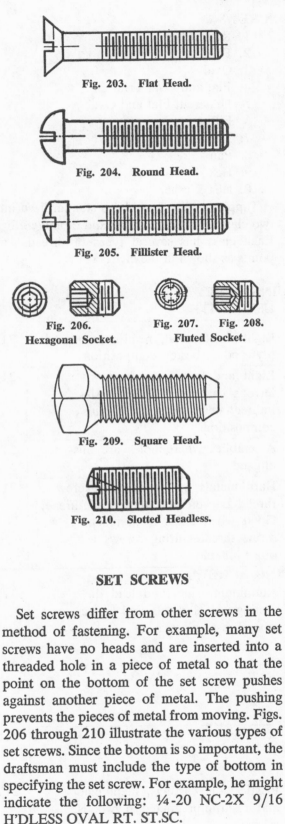

Fig. 203. Flat Head.

Fig. 204. Round Head.

Fig. 205. Fillister Head.

Fig. 206. Hexagonal Socket.

Fig. 207. Fig. 208. Fluted Socket.

Fig. 209. Square Head.

Fig. 210. Slotted Headless.

SET SCREWS

Set screws differ from other screws in the method of fastening. For example, many set screws have no heads and are inserted into a threaded hole in a piece of metal so that the point on the bottom of the set screw pushes against another piece of metal. The pushing prevents the pieces of metal from moving. Figs. 206 through 210 illustrate the various types of set screws. Since the bottom is so important, the draftsman must include the type of bottom in specifying the set screw. For example, he might indicate the following: ¼-20 NC-2X 9/16 H'DLESS OVAL RT. ST.SC.

SHOULDER SCREWS

Shoulder screws, shown in Figs. 211 and 212, are used as pivots for cams and come in nominal sizes of ¼" to 1¼". Table III provides the standard dimensions. Shoulder screws are commonly called **stripper bolts.**

Fig. 211.
Hexagonal.

Fig. 212.
Fluted.

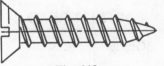

Fig. 213.

TYPE BP

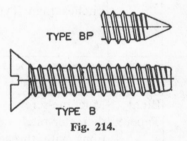

TYPE B
Fig. 214.

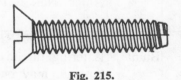

Fig. 215.

Fig. 216.

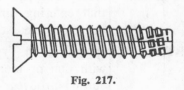

Fig. 217.

TAPPING SCREWS

Since a **tapping screw** is driven into two pieces of material to fasten the material together, it creates internal threads in the material. First a hole is driven into the material. The diameter of this hole should be smaller than the diameter of the screw. As the tapping screw goes into the hole it makes the threads in the material. Therefore the need to tap a hole separately is eliminated. Furthermore, some materials, such as asbestos, plywood, and plastics, do not lend themselves to easy tapping with regular taps and dies. Tapping screws are a great help in working with these types of material.

Tapping screws with the following ten types of heads can be obtained:

1. Round
2. Flat
3. Oval
4. Flat and Oval Trim
5. Undercut Flat and Oval
6. Fillister
7. Truss
8. Pan
9. Hex
10. Hex Washer

Tapping screws have been standardized into two classes, **thread forming** and **thread cutting**. Each class has several types of threads and points as shown in the charts below.

THREAD-FORMING SCREWS

TYPE	POINT	THREAD	USED WITH	FIG.
A	Gimlet	Spaced	Light sheet metal, resin-impregnated plywood, asbestos compositions	213
B	Blunt	Spaced with finer pitches than Type A	Light and heavy metals. Non-ferrous castings, plastics, resin-impregnated plywood, asbestos compositions	214
BP	Cone	Same as Type B	Assemblies where holes are mis-aligned	214
C	Blunt Tapered	Machine-screw diameter-pitch combinations with threads like Amer. Nat. form	Hard metals, where machine-screw thread is preferable to spaced thread. Good when chips from machine-screw thread-cutting screws are objectionable	215

THREAD-CUTTING SCREWS

TYPE	POINT	THREAD	USED WITH	FIG.
F	Blunt	Entering threads are tapered, may be complete or incomplete. Have one or more cutting edges, and chip cavities. All other threads like machine-screw threads	Aluminum, zinc, and lead die castings, steel sheets and shapes, cast iron, brass, plastics	216
G,D,T	Blunt	Same as Type F, except that entering threads are incomplete only	Same as Type F	216
BF,BG,BT	Blunt	Spaced, as in Type B, with one or more cutting grooves	Plastics, die castings, metal-clad and resin-impregnated plywoods, asbestos	217

METALLIC DRIVE SCREWS

For making permanent fastenings in metal and plastics, metallic drive screws are used. They are forced into the material by pressure. The screws are multiple-threaded with a large helix angle and have a pilot. Fig. 218 illustrates metallic drive screws.

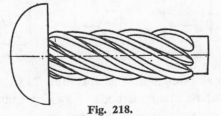

Fig. 218.

BOLTS AND NUTS

Bolts. A bolt has a head on one end of a straight piece of steel and threads on the other end. To fasten two pieces of metal, simply slip the bolt through matched holes in the metal pieces and screw a nut on the threaded end. All bolts and nuts have been standardized as **regular** or **heavy.** Regular bolts and nuts are used for general purposes, and heavy ones are used when more bolt-head surface pressing down on the fastened metal is needed. Heavy bolts and nuts are larger and heavier than regular bolts and nuts.

Bolt heads may be either **square** or **hexagonal.** Square heads, shown in Fig. 219, come only in the regular series. Hexagonal heads, as shown in Fig. 220, may be obtained in both the regular and heavy series. Square heads can be obtained only as unfinished because they can be used only when the work does not require a great deal of accuracy. Heavy and regular heads are obtainable as unfinished and semifinished. Fig. 221 shows a regular semifinished hexagon bolt, and Fig. 222 shows a regular finished hexagon bolt. The draftsman may also be called upon to work with heavy semifinished and finished hexagon bolts. Semifinished bolt heads are machined and have a more accurate surface than unfinished bolt heads.

The draftsman must indicate that he wants a heavy bolt, but he does not have to indicate that he wants a regular bolt.

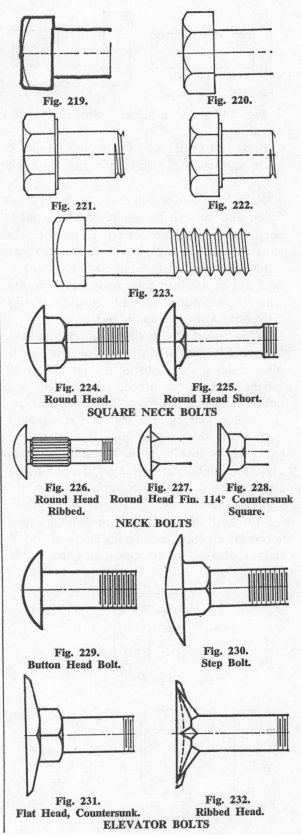

Fig. 219.

Fig. 220.

Fig. 221.

Fig. 222.

Fig. 223.

Fig. 224.
Round Head.

Fig. 225.
Round Head Short.

SQUARE NECK BOLTS

Fig. 226.
Round Head
Ribbed.

Fig. 227.
Round Head Fin.

Fig. 228.
114° Countersunk
Square.

NECK BOLTS

Fig. 229.
Button Head Bolt.

Fig. 230.
Step Bolt.

Fig. 231.
Flat Head, Countersunk.

Fig. 232.
Ribbed Head.

ELEVATOR BOLTS

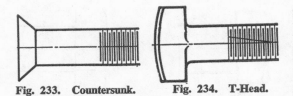

Fig. 233. Countersunk.   Fig. 234. T-Head.

Fig. 223 shows a lag bolt which is used in wood.

Figs. 224 through 234 show other types of bolts with which the draftsman must be familiar.

Nuts. Nuts conform in shape to the heads of bolts. The nut can be **square** or **hexagonal** in shape and **regular** or **heavy** in weight. The square nut comes in regular and heavy weights, but in unfinished surface only, while the hexagonal nut comes in regular and heavy weights, with both unfinished and finished surfaces. Fig. 235 shows a **regular square nut.**

When bolted objects vibrate or are pounded, nuts tend to loosen. **Hexagon jam nuts stay in place.** They are available in regular and heavy weights and in unfinished, semifinished, and finished surfaces. See Figs. 236 through 238.

Another locking nut, called the **hexagon slotted and thick-slotted nut,** is used with a **cotter key** or **locking wire.** This nut, shown in Fig. 239, can be obtained in regular and heavy weights with a semifinished surface, and with a hexagonal head only.

Figs. 240 through 242 illustrate **hexagon finished thick nuts, hexagon finished castle nuts, and machine-screw and stove-bolt nuts.**

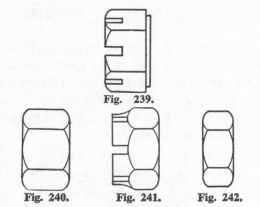

Fig. 239.

Fig. 240.   Fig. 241.   Fig. 242.

Lock Nuts and Washers. Figs. 243 through 249 show other methods of locking nuts. Fig. 250 shows standard washers which are used to provide a wide area for the bolt head.

Fig. 251B shows the various uses of lock nuts and washers.

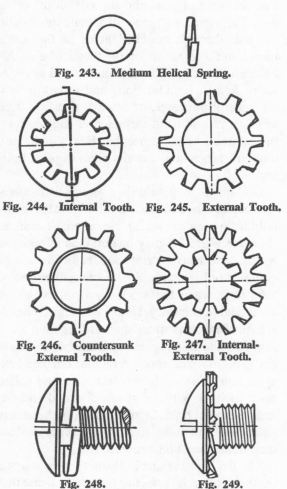

Fig. 243. Medium Helical Spring.

Fig. 244. Internal Tooth.   Fig. 245. External Tooth.

Fig. 246. Countersunk   Fig. 247. Internal-
External Tooth.   External Tooth.

Fig. 248.   Fig. 249.
Lock Washer Assemblies.

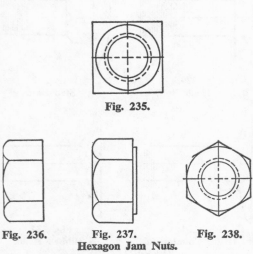

Fig. 235.

Fig. 236.   Fig. 237.   Fig. 238.
           Hexagon Jam Nuts.

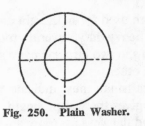

Fig. 250. Plain Washer.

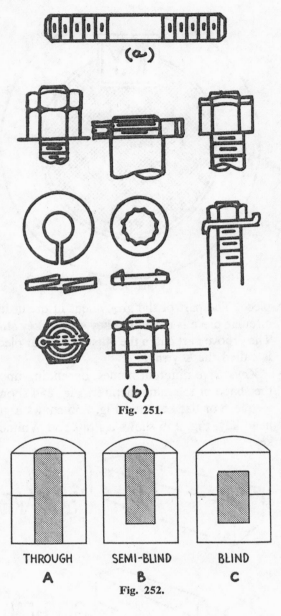

(a)

(b)

Fig. 251.

STUDS

A stud is similar to a bolt. It has **no head,** however, and has threads on each end as shown in Fig. 251A. Of the two pieces of metal to be fastened with a stud bolt, the deeper one has a tapped hole. The nearer piece of metal has only a drilled hole. Screw the stud into the tapped hole. The end of the stud going into the tapped hole is called the **stud end.** The other threaded end of the stud sticks out. A nut, called the **nut end,** screws on to the end which sticks out. The American Standards Association has not standardized studs. The draftsman must determine and indicate on a detailed drawing how much of a threaded length he needs, what diameter stud he needs, how long a stud is required, and what type of thread is needed.

DOWELS, TAPERS, AND KEYS

Bolted or screwed machine parts may shift or loosen. Dowels, tapers, and keys help prevent loosening or shifting. The draftsman may have to draw or specify them.

A dowel pin, made of high-carbon steel, may be either a straight rod or tapered. The straight pin varies in diameter from 1/16″ to 2″ and ranges from ¼″ to 2″ in length. It fits into holes in the two machine parts. When the two holes go right through the two parts, as in Fig. 252A, the dowel is said to be a **through** type. When the hole in one of the metal parts goes partly through, the dowel is said to be a **semiblind** type, as shown in Fig. 252B.

The smaller hole, continuing from the partly drilled one, provides an air vent. The air vent may sometimes appear perpendicular to the partly drilled hole.

A seldom-used type of dowel, called the **blind type,** is shown in Fig. 252C.

| THROUGH A | SEMI-BLIND B | BLIND C |

Fig. 252.

KEYS

Keys prevent rotation of wheels or gears on shafts and generally where two circular parts must move together. The key itself is a piece of metal, round, square, or semicircular, that sits between the two circular pieces of metal, as shown in Fig. 253. The key sits in a groove cut to match the shape of the key. The groove is cut partly in the inside circular metal piece to be held together, and partly in the outside circular

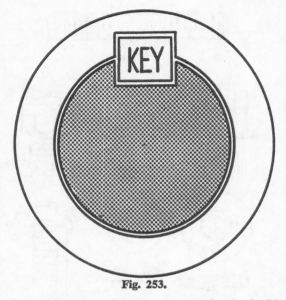

Fig. 253.

piece. The part of the groove cut in the inside circular piece is called the **key seat** or **key slot.** The groove part cut in the outside circular piece is called the **keyway.**

Keys have different names, depending upon the shape or the manufacturer. Fig. 254 shows a square or flat key, and Fig. 255 shows a gib head key. Fig. 256 shows a Pratt and Whitney key.

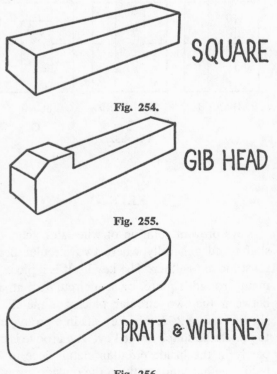

SQUARE

Fig. 254.

GIB HEAD

Fig. 255.

PRATT & WHITNEY

Fig. 256.

Keys are used for both **light** and **heavy** duty. For **very heavy duty,** several grooves or **splines** are cut in the two circular parts so that they fit into each other.

The draftsman must indicate the keys to be used. He gives the width, length, and height of square and flat keys as, for example, ¼ Square Key 2LG, or ⅜ × ¼ Flat Key 2LG. He describes tapered keys by giving length, width, and height at the large end as, for example, ½ × ½ × 2 Square Plain Taper Key, or ⅜ × ¼ × 2LG Flat Plain Taper Key. If the key is a gib head, he includes the name in the note. A rounded key must show diameter, thickness, and length.

The draftsman must indicate the key required by number or letter. Some companies designate their keys by number and others use letters.

Keys must also be dimensioned. If one key is required, nominal dimensions are given. If many keys must be made, limit dimensions should be included. For the proper way to dimension keys made by a specific manufacturer, the draftsman should consult the company catalogue.

DRAWING SCREW THREADS

The American Standards Association has standardized three methods of representing screw threads. The draftsman may show the threads in **detail,** as in Fig. 257, **schematically,** as in Fig. 258, or in a **simplified version,** as in Fig. 259. The simplified version may help save drafting time, the schematic drawing may be used for mere illustration, and might appear with the detailed version in one drawing, as shown in Fig. 260. (See A.S.A. Y. 14.6.)

The detailed drawing shows the threads as they actually slant. The ends of the threads, the crests and roots, appear "V"-shaped.

The schematic drawing shows the threads as single, heavy lines with thinner lines between. The heavy lines represent the threads to the roots, and the thin lines show the threads to the pitch. The lines may be perpendicular to the center line of the screw or slanted to show how the threads run.

The simplified representation shows the pitch diameter as the outline of the screw and a dotted line as the root diameter. See Fig. 259. The simplified version should be used to show hidden internal threads.

When the draftsman is required to give a more detailed presentation of a thread, a drawing, such as the one shown in Fig. 257, may be used. Fig. 185 illustrates the words with which the draftsman must be familiar when working with screws.

Fig. 259.

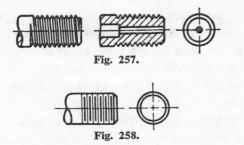

Fig. 257.

Fig. 258.

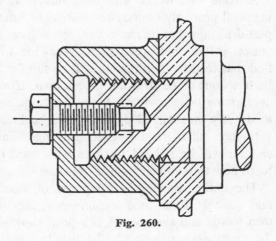

Fig. 260.

CHAPTER TEN

READING A DRAWING

Anyone who works with mechanical drawings will probably be required to read or interpret drawings that he has never seen before. A person called on to read such a drawing will find that reading the drawing will be simpler if he has done some drawing on his own. However, many people who may never be required to do any mechanical drawing may be faced with the task of reading such drawings. This chapter will be helpful to those who want to learn how to read mechanical drawings.

The entire drawing is composed of **words** and **symbols.** Each industry, however, uses its own words and symbols. A blueprint showing the construction of a desk will naturally contain words and symbols that are entirely different from a drawing showing how steam goes through a building. In learning to read a drawing the reader must understand what each technical word means and what each symbol represents. The reader must then study the blueprint to find how these symbols are related to each other. Notes on the drawing will provide the reader with information about certain specific parts. A separate list of instructions, which gives the exact type and size of certain parts, the method of fastening parts, finishing techniques, and any other specific requirements that the designer considers essential, is found on a separate sheet called a **specification sheet.**

In studying a drawing the reader must consult the specifications, commonly called "specs." and all the notes on the drawing. A typical specification sheet is shown in Fig. 287, Chapter Eleven.

The person studying the blueprint must visualize the object as it would look in **three dimensions.** The three views, the sections, and the details show the object and its parts in two dimensions. A rough freehand three-dimensional sketch often helps to clarify the reader's concep-tion of the object. If you find freehand sketching difficult, practice with a triangle and a T square. A little practice in freehand sketching will enable you to develop the necessary control over your hand.

If the reader is able to visualize the entire object, the details will fall more clearly into place. If the reader wants to comprehend the total picture, he must see the three views in relation to each other. The information lacking in any one view will be found in either or both of the other views.

If the drawing portrays a very large object, it will be necessary to learn first what the whole object looks like. Separate parts can then be studied more effectively.

Architects always draw a perspective view of a proposed building. This makes it easier for the blueprint reader to understand the construction details. The reader must know the words that are used in architectural drawings, the symbols for the various materials, the methods of fastening parts and the general construction methods used when working with wood, brick, stone, or steel structures.

A list of the more common terms used in architectural drawings will be found at the end of this chapter. The blueprint reader will find no advantage in merely memorizing the definitions. He must be able to visualize the location of each part of the building. The accompanying illustrations should help him to accomplish this. The symbols for various materials will be found in Chapter Eleven.

The method used to assemble the various parts of the structure depend upon the materials being used. Wood, brick, concrete, and steel parts require different fastening procedures. Wood and cement are nailed. Brick and concrete blocks are mortared. Solid concrete is poured, and steel parts are welded or riveted. Directions and standards for fastening usually appear in the specifications.

Since every building has its own construction details, the blueprint reader must become acquainted with the details of the building that he is concerned with so that he will know how the particular building is to be assembled. The reader must examine every detail. He must learn the size of each piece and where each piece must be placed. Most boards are rectangular and the reader must determine whether the narrow dimension is vertical or horizontal. He must know the size of the footing, the foundation material, the grade elevation, the floor-joist spacing, the direction of the subflooring boards, the size and spacing of the studs, the type of building paper, the sheathing, the kind and size of the shingles, the type and location of the bracing, the shape of the roof, and all the details of the framing. The windows, doors, and stairs have their own construction details which must be studied.

Of course, standard construction methods have been developed for many parts of a building. Besides, doors, windows, cabinets, and other items may be purchased in finished form and merely installed. However, the blueprint reader must make himself aware of any deviation from the standards. The door may be a purchased item, but certain variations in its frame or in its installation may be necessary. It is the variations which the reader must look for.

ESSENTIAL VOCABULARY

The following section contains several groups of words, divided by categories, with which the reader of blueprints must be acquainted. Knowing and understanding them will help the reader to understand many different kinds of architectural drawings. The words are divided into different categories by subject matter.

"FOUNDATION" WORDS

Footing	Supports foundations. Distributes load of foundations. See Fig. 261.
Foundation	Supports structure. Usually made of stone, concrete, or brick. Prevents wood in frame buildings from being rotted. Provides a solid base for brick structures. See Fig. 261.
Pier	A vertical part of wood, brick, or concrete, fastened to a sill to support the joists or to keep a temporary structure off the ground. See Fig. 262.
Grade Lines or Lot Grade	Natural surface of ground around a building. It may be the top of a foundation wall. **Finished grade** is the surface line after the ground has been leveled and is usually determined by the building owner. See Fig. 263.
Frost Line	Maximum depth to which ground freezes. Footing usually exists **below** the frost line.
Sill	Horizontal piece that rests on foundation or pier to support superstructure. See Fig. 264. See "Window" Words for other types of sill.

"DOOR" WORDS

Door Size	The door size must be given and shown in the drawing. See Fig. 265.
Head	One piece in the frame structure across the top of the door opening. See Fig. 266.
Angle Iron	A piece of steel in the shape of an angle that is embedded or placed in headers to support the structure above the door openings. See Fig. 267.

Casing or Trim A board or molding used to cover the edge of the plastering around door openings. See Fig. 268.

Jamb The sidepost of a doorway. See Fig. 268.

Door Stile One of the vertical, outside edge pieces of the frame of a panel door.

Threshold A piece of beveled wood, stone, or metal under a door. Also called a **saddle.**

"STAIR" WORDS

Shoe A molding fastened to the stringer, into which the banisters are set.

Stringer or String Board One of the pieces supporting the treads and risers of a flight of stairs. A wall stringer is attached to a wall. See Fig. 269.

Riser The vertical piece of wood between the two horizontal treads. It forms the front of the step.

Tread The horizontal part of the step.

Fire Stop Blocks nailed between studs or joints, often under stairs, to prevent drafts in case of fire.

Nosing The rounded part of a tread that projects beyond the riser. Return nosing is the continuation of nosing at right angles.

Soffit The underside of a flight of stairs.

"ROOF" WORDS

Truss A framework to support a roof.

Run The horizontal distance of one sloping part of a roof. See Fig. 270

Rise The perpendicular distance of a roof. See Fig. 270.

Span The total horizontal distance of a roof. See Fig. 270.

Pitch The slope or angle of the roof.

Rafter One of the pieces of wood supporting the roof shingles or covering. A **common** rafter is one of the **main** rafters. A **jack** rafter is at the same pitch as common rafters, but is shorter in length and fits on a **hip** or **valley** rafter. A **hip** rafter is the corner rafter. A **valley** rafter is the main rafter in a **valley** roof. See Fig. 270.

Eave An eave is the lower edge of a roof.

Water Table In brick or stone buildings the water table is an extension of the eaves to form a water drop.

Cornice Part of the roof which projects beyond the building.

Gable End The end of a gable roof.

Purlin A timber supporting the rafters of a roof. It is located between the seat and the comb.

Roof Boards Boards fastened to the rafters to serve as a base for the roof covering. Roof boards are also **roof sheeting** or **sheathing boards.** See Fig. 264.

Collar Beam or Tie Beam A horizontal piece of wood that connects two opposite rafters. A collar-beam roof is one in which all the rafters are held together by collar beams. See Fig. 270.

Frieze A board forming a band immediately below the cornice. See Fig. 268.

Plancher The bottom board of a cornice. It is also called a **soffit.** See Fig. 268.

Crown Molding	The molding directly under the shingles of a cornice. See Fig. 268.
Bed Molding	A molding smaller than the crown molding and used in the angle between the plancher and the frieze.
Roofers	Tongue-and-grooved boards used as a roof covering over rafters, as a base for roofing material. Usually made of inferior material.
Lookouts	Projecting ends of timber onto which finishing boards called **fascia** are nailed. See Fig. 271.
Roofing Felts or Paper	Felt, asphalt, and gravel in tarred-paper form used as roofing material.
Flashing	Pieces of tin or copper that are used around dormers, chimneys, or gutters to prevent leakage.
Roof Shingles	Asphalt, slate, tile wood, or asbestos pieces used as a roof covering. See Fig. 268.

"WALL" AND "FLOOR" WORDS

Wall Types	a. **Bearing Wall**—Supports loads other than its own weight. b. **Curtain Wall**—Supports no other load but its own weight. c. **Party Wall**—Wall dividing two adjacent pieces of property, or two adjacent apartments.
Floor Joist	Attached perpendicular to the sill. Usually 12 inches or 16 inches on centers. See Fig. 264.
Bridging	Pieces put across or between joists to stiffen them and distribute the load.
Subflooring or Rough Flooring	Attached to the floor joists to serve as a base for the finished floor. See Fig. 264.
Finished Floor or Finish Floor	Surface attached to a subfloor or rough floor. The finished floor line on an elevation shows the top of the finished floor. See Fig. 264.
Shoe	Piece put across subflooring to serve as a base for studs. It may also be the molding between the floor and baseboard.
Stud	Vertical pieces 2×3, 2×4, or 2×6 inches that run across the height of a building. They are attached to a shoe at 16-inch intervals. These **common** studs serve as a frame to which the walls and partitions are attached. **Jack** studs are shorter, vertical pieces that are used when cutouts for windows and ventilation, etc., are required. **Trimmer** studs are shorter than common studs and are used around floor openings. See Fig. 264.
Beam	A horizontal piece supporting the weight of the structural frame. The beam may be made of wood, concrete, or steel.
Girder	A horizontal beam to span an opening or carry weight. See Fig. 272.
Post	A vertical piece of wood used for supporting a frame. See Fig. 272.
Header	A horizontal piece of wood in the frame that goes across the top of a rough window opening.
Stretcher	A method of laying bricks so that the long edge of the brick is visible.

Brace　　A piece nailed between studs for stiffening.

Sheathing　　Gypsum, fiberboard, or wood boards, usually 1'×6' or 1'×8', nailed to studs to provide a base for a finished wall or to serve as a wall or as insulation. See Fig. 264.

Batten　　A piece of wood nailed across other pieces of wood to hold them together or to cover the crack between two pieces.

Lath　　A narrow strip of wood, about 4 feet long, nailed to studs to support plaster. Lath may be made of metal latticework.

Ribbon　　A board attached to the studs to support the second-floor joists. See Fig. 264.

Furring　　Vertical strips of wood or metal attached to the inside surface of a wall or ceiling to provide a base for attaching lath, boards, or plastic and to create an air space between outside and inside walls.

Building Paper　　Paper put between the wall and siding or between the subfloor and finished floor. See Fig. 264.

Shingle or Siding　　A piece of wood 4 inches wide and 16, 18, or 24 inches long, which tapers to a thin point. Shingles placed over each other in certain ways on the sides or roof of a wooden building serve as the covering or outside surface. See Fig. 264.

Course　　A layer of shingles or bricks, running the length of a wall. The method of placing bricks next to each other determines the name of the course, such as soldier course, stretcher course, etc.

Anchor Bolt　　A bolt used to fasten sills and wall plates to foundation walls. See Fig. 262.

Wall Plate　　A horizontal wood piece set on a wall, often used to support the rafters. See Fig. 264.

Drain Tile　　Tile around the foundation wall below the basement level. It is used to drain off water from the soil before it gets into the basement.

Corbel　　A block of timber with beveled ends resting on a post to support beams.

Millwork　　Standard parts for sash, doors, and windows as well as the finished doors, windows, and sash that are made in factories and mills.

Plank　　Wood pieces that are 2 inches thick and 4 inches wide. They are actually 1⅝" ×3⅝".

Shoring　　Supporting anything with shores. Shores are wood pieces that resist pressure. Shoring is used under or against weak or bulging walls.

"WINDOW" WORDS

Window Types　　See Chapter Eleven, Architectural Drawing.

Fenestration　　The arrangement of windows in a building.

Window Frame　　The wood shape into which the sash are fitted. Frames may be of either the skeleton or box type.

Window Head　　A piece that goes across the top of a window. It is also called a lintel. See Fig. 266.

Blind Casing　　The inside piece of a window frame that forms a rough casing. See Fig. 266.

Interior Casing
The trim around windows on the inside of a building. See Fig. 268.

Parting Strip or Stop
The piece that separates the upper and lower sash. See Fig. 266.

Window Sash
Units of a window. Sash is used as both singular and plural. A double-hung window has two sash. See Fig. 266.

Rails
Top, Bottom, Meeting—Horizontal pieces in the window sash.

Muntin or Mullion
A small box separating window panes.

Stool
A horizontal piece of window finish that forms a kind of stool for the side casings and conceals the window sill. See Fig. 266.

Sill
An underpiece on which a window rests. A **lug** sill in a brick house is a sill with a lug on each end for anchorage. A **slip** sill in a brick building is slipped under the window frame after the frame has been set and is held in place only by mortar joints. See Fig. 266.

Calking
Filling up cracks for making a waterproof or airtight window. See Fig. 266.

Apron
The piece below the stool in a window casing. See Fig. 266.

Drop Cap
A piece of wood or molding that juts out over the top of a window to let the water run off.

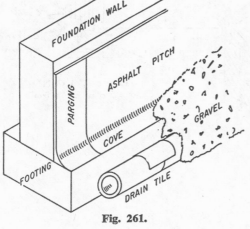

Fig. 261.

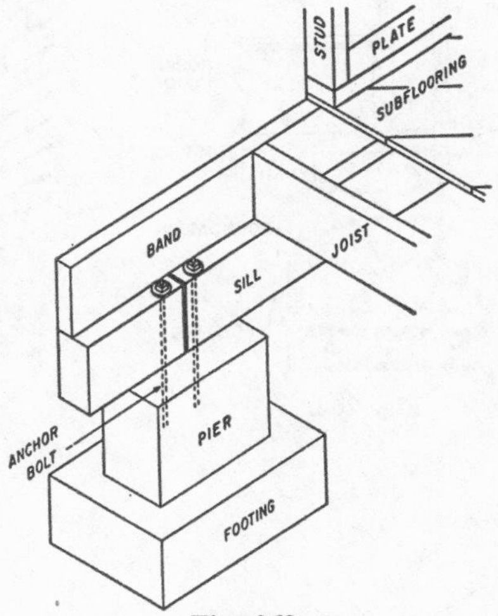

Fig. 262.

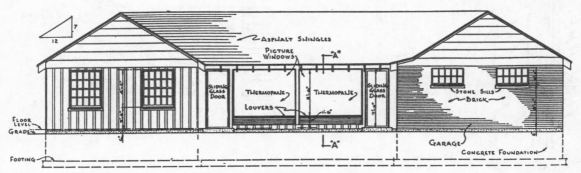

Fig. 263.

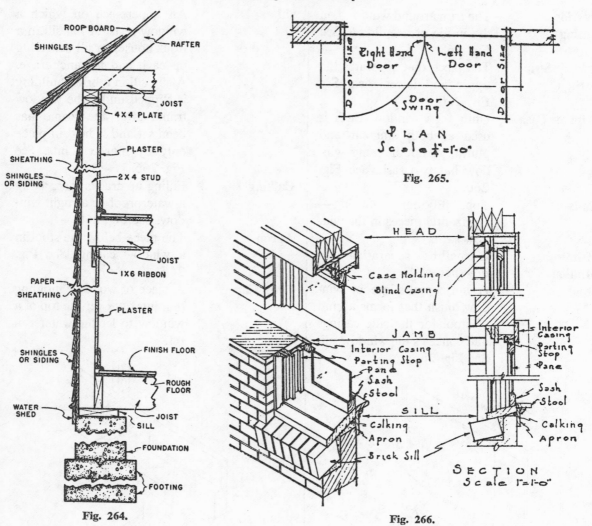

Fig. 264.

Fig. 265.

Fig. 266.

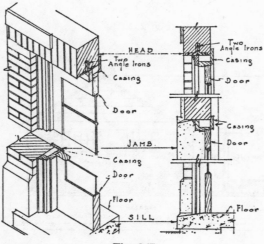

Fig. 267.

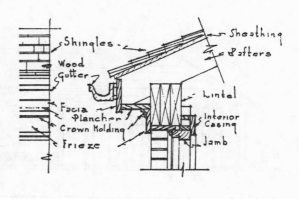

Fig. 268.

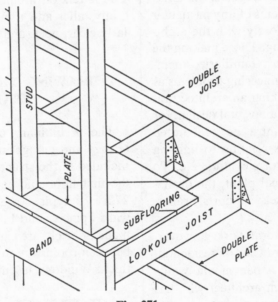

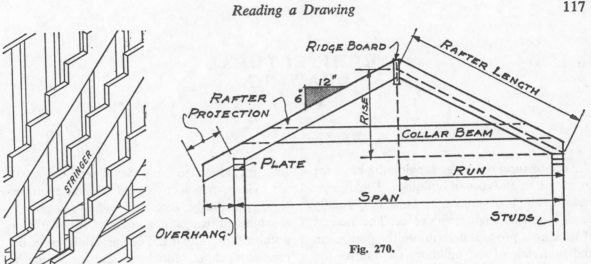

Fig. 269.

Fig. 270.

Fig. 271.

Arrangement and nailing for
built-up beams or girders.

Fig. 272.

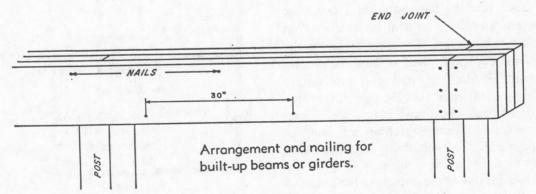

CHAPTER ELEVEN

ARCHITECTURAL DRAWING

Architectural drawings provide the basis for constructing all types of buildings. The designs and drawings of industrial buildings usually come out of an engineering office. The method of making a product determines the plan, size, and materials of the building. Of course, the climate, temperature, type of soil, and cost of materials, labor, and land also influence the design of industrial buildings. Engineers familiar with the engineering problems of any particular industrial process work closely with the architect in planning the building. Those making the architectural drawings must familiarize themselves with some of the engineering problems as well as the methods of making an architectural drawing. In order to draw a hospital design, for instance, the draftsman must be acquainted with the specific problems relating to lighting operating rooms, the size of elevators, etc. To draw a design of a chemical plant, the draftsman must know about vessels, supports, cranes, handrails, etc.

A good deal of architectural work, however, involves the design and drawing of residential and other small buildings. A person who wants a home designed goes to an architect who has designed houses similar to the type he wants. He tells the architect how much he wants to spend, how much land he has, how big a house he wants, and all other pertinent information that might affect the design. The architect knows the cost of materials, of labor, of land, of his fee. With all this data he makes a preliminary sketch, probably freehand. He gives this sketch to a draftsman, who draws it to **scale.** Sometimes several alternate preliminary sketches may be made. The preliminary sketch consists mainly of the plan views of each floor, and a rough **perspective** of the outside of the house.

All of the principles of orthographic projection, including the three views, sections, auxiliary views, perspective, and isometric projection, previously discussed with relation to machine drawings are equally applicable to architectural drawing. The basic differences between machine drawings and architectural drawings will be found in certain dimensioning practices, general and local notes, and lettering. The letters on architectural drawings are generally taller and thinner and must be particularly clear, neat, and legible.

DRAWING THE PRELIMINARY PLAN VIEW

The preliminary plan view consists of the terrace and garage, if either or both are to be included in the structure, and all the rooms on the first floor and each succeeding floor. Fig. 273 is a typical preliminary plan view. The walls appear solid black. The outside walls, if made of brick, are eight inches thick. The walls between rooms are much less than eight inches thick. Windows in outside walls may appear in

Fig. 273. Preliminary Plan.

any of the ways shown in Fig. 274. Space is allowed for doors in inside walls, and an arc is drawn in that space to show which way the door swings. See Fig. 274. Each room is labeled and sometimes the over-all dimensions are given. Stairs, if any, are included and the drawing indicates whether they lead up or down, as shown in Fig. 275. Fireplaces, bathroom fixtures, walls, and closets also appear in the preliminary sketches. Thin lines are used for all outlines, except the walls. The scale may vary depending upon the size of the house.

PRESENTATION DRAWINGS

A more elaborate drawing, showing the trees, shrubbery, and other landscaping features, and any special materials, such as flagstone, is called a **presentation drawing.** Water colors, tempera, pencil, or oils may be used. The elevation, drawn in perspective, has human figures to show relative proportion. Sheets with vanishing lines printed on them can be placed under the transparent drawing paper or linen to help draw the elevation in a more accurate perspective.

The presentation drawing, with a plan view of the first floor, is often drawn by a commercial artist or one specializing in such drawing and not by the architectural draftsman.

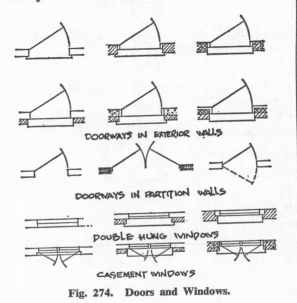

DOORWAYS IN EXTERIOR WALLS

DOORWAYS IN PARTITION WALLS

DOUBLE HUNG WINDOWS

CASEMENT WINDOWS

Fig. 274. Doors and Windows.

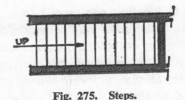

Fig. 275. Steps.

MODELS

In order to give a realistic idea of the finished building, an architect may submit a model of the house to his client. Models can be made of simple materials and quickly, if the building is not too large or complicated. Model making, however, is generally left to people who do nothing else but build models. Architects have these models made when they submit competitive bids. Models are made strictly to scale from drawings supplied by the architect.

WORKING DRAWINGS

Besides contributing the design, the architect must produce drawings so that the construction company will be able to erect the building. These **working drawings** must include all the information and construction details necessary to build the object. Working drawings consist of a **plot plan, floor plans, elevations, sections,** and **enlarged detail drawings.** With these drawings the architect presents a list of written instructions called **specifications.**

PLOT PLAN

The **plot plan,** also known as a **site plan,** shows the finished house in **plan view,** as it will look in relation to the surrounding area. The drawing includes the landscaping, if any, around the house, the location of the streets, and the water, gas, electricity, and sewage-disposal systems. The plot or site plan locates the exact area of the house and tells the builder just where to erect it. All drawings must be submitted to the proper municipal authorities for approval. The plot or site plan must show that the building location conforms to the zoning regulations. Some areas may be restricted as to size and design, others may require buildings to be a certain distance from the curb or roadway. Chemical plants, for instance, must be constructed so that they do not endanger the public with fumes or chemicals. Fig. 276 illustrates a plot or site plan.

FLOOR PLANS

A plan view of each floor must be drawn by the draftsman. The **floor plan** shown in Fig. 277 locates windows, doors, walls, stairways, bathroom equipment, and closets. Floor plans often locate elevators, fireplaces, electrical outlets, radiators, and other nonmovable features. A floor plan also indicates the thickness of the inside and outside walls, and the materials of which they are made.

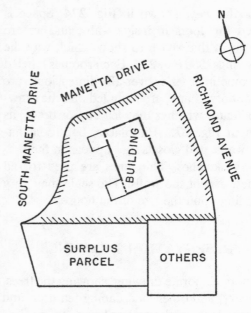

Fig. 276. Site Plan.

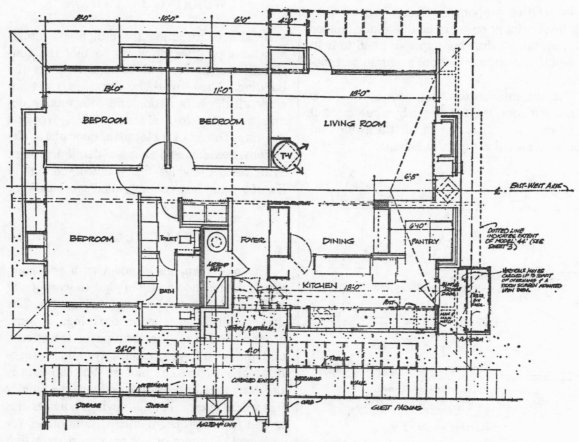

Fig. 277. Floor Plan.

DRAWING THE FLOOR PLAN— STANDARD GRID

The American Standards Association has attempted to establish some means by which all the materials that are used in a building will have a definite size relationship. This is shown in Fig. 278. All building materials should be made **4 inches in length, width, and depth, or in multiples of 4 inches.** A piece of wood, for instance, would be available in lengths and widths of **4, 8, or 12 inches.** The design of a house would also be based on a **4-inch grid.** A wall would be **8 inches,** not 6 inches, thick. Doors would be **2′-0″, 2′-4″, 2′-8″ wide—** and so on.

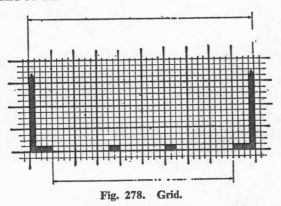

Fig. 278. Grid.

Many manufacturers do not abide by the standards. Their building products may be 6, 9, or 10 inches in length, width, or height. Many architects will specify dimensions that are not multiples of 4 inches.

All this affects the drawing. A graph paper, drawn on the 4-inch-grid basis, is used under the drawing sheet. The graph paper makes it easier to locate walls, doors, and other parts of the house at grid dimensions. The plan should be arranged so that the front entrance of the house appears toward the bottom of the sheet. The outside walls should be located first. The walls could coincide with the grid lines, as in Fig. 279A, or they could be centered from the grid lines, as in Fig. 279B. Dimensions can be more easily drawn and recognized in this way. In elevations, give floor, ceiling, and wall openings on or from grid lines. When grid dimensions are used, the standard dimension line,

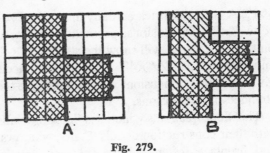

Fig. 279.

with arrows at each end, is used. Grid lines generally do not appear on a professional drawing. A sheet of paper with horizontal and vertical grid lines is placed under the drawing as a guide. When nongrid dimensions appear, use a dimension line with an arrow at one end and a dot at the other end, as shown in Fig. 280. Fig. 277 indicates the proper way to denote walls. Windows should be located in the walls and drawn with the appropriate symbol, as illustrated in Fig. 274. The addition of further details of wall and window construction will depend on the type of wall and window material used. The draftsman must learn the various combinations of wall and window types and the ways of constructing them. Frequently one part of a wall and window appears in an enlarged detail.

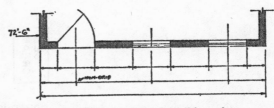

Fig. 280. Grid and Non-Grid Dimensions.

The face of one outside wall is located from some definite point outside the building, such as from the edge of the plot. Dimensions are given to the outside or face of exterior walls, to the outside face of studs in exterior walls, and to other places, as indicated in Fig. 276.

Enlarged detail drawings include dimensions of the bricks, studs, doors, walls, and of every other important construction material. These dimensions do not always tell us the true size of the material. In a 3-foot doorway, the door itself is not 3 feet. A 2″×4″ stud is only 1⅝″×

3⅝″. A common brick actually measures 2¼″ ×3¾″×8″. Dimensions, particularly if based on the 4-inch grid, will show the stud as 2″×4″ and the brick as 2½″×4″×8″. Some architects give the actual dimensions while others give the rounded-off dimensions.

The floor plan also includes interior walls with their construction details. Stairs, windows, and fireplaces often require enlarged details. See Fig. 281.

COMMERCIAL BUILDINGS— FLOOR PLANS

Commercial buildings usually have columns or beams made of steel or reinforced concrete. These steel and reinforced concrete pieces support the walls and floors, and appear in structural floor plans. The concrete foundation and special concrete work appear in separate concrete-floor plans. The architectural floor plan for each floor shows the exterior walls, interior walls, windows, doors, stairs, and roof. The floor plan gives location and size dimensions. Construction information appears in enlarged details and sections. See Figs. 282 and 283.

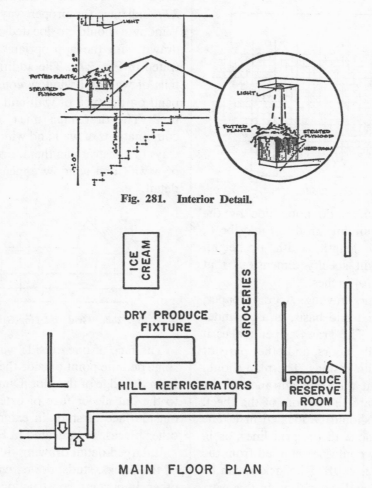

Fig. 281. Interior Detail.

MAIN FLOOR PLAN

Fig. 282. Floor Plan.

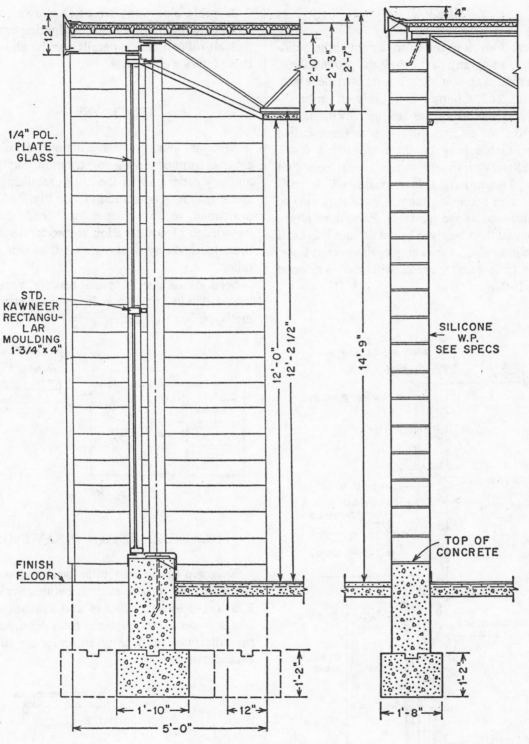

Fig. 283. Detail.

ELEVATIONS

Elevations show the entire height of the building, including any part below the ground floor. Each side of the building has its own elevation. See Fig. 263, Chapter Ten. Only elevation dimensions should appear on the elevation. The ground floor or first floor is sometimes called **grade.** Grade may be dimensioned as 0'-0", and all elevations are dimensioned from that point. Dimensions **below grade** will have a minus sign preceding them. Grade may also be dimensioned above sea level. Elevations **above grade** will then be given based on that height. If grade is 126'-0", for example, the second floor, if it is 15 feet above grade, will be dimensioned as 141'-0".

Separate elevations are often drawn for exteriors and interiors. Furnaces, lighting fixtures, and all other interior details can be shown in these interior elevations.

SECTIONS

Sections portray construction details of closets, furnaces, wall construction, skylights, and any other part of the house requiring special methods of construction. See Fig. 284. The draftsman, to do a proper job, should have a knowledge of construction methods. He learns these methods by working as a tracer or a detailer.

Sections of special parts usually have enlarged details showing individual construction methods. See Fig. 285.

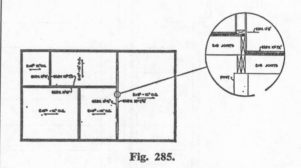

Fig. 285.

ENLARGED DETAIL DRAWINGS

Enlarged details may often be drawn on one sheet devoted only to such drawings. **Scales of 1 inch, 1½ inches, 2 inches, and 3 inches to the foot are used.** If the scale for each detail varies, then the scale must be given under the title of each detail.

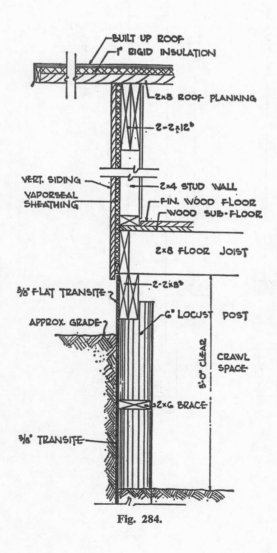

Fig. 284.

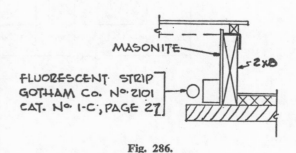

Fig. 286.

All materials must be noted. If standard items, such as a hang rod in a closet, are used, then the manufacturer, catalogue number, and page and name of the item must be included. See Fig. 286.

Vendors prints must also be included in drawings. These prints come from the manufacturers of items ordered by the architect and include such items as boilers, bathroom equipment, and lighting fixtures.

An enlarged detail may have to be drawn near a section, plan, or elevation to show some special construction. A circle should be drawn around this special area in the section, plan, or elevation. A line is then drawn from this circle to a much larger circle near the section,

plan, or elevation. The enlarged detail appears in this larger circle. See Fig. 285.

SPECIFICATIONS

A list of general and special instructions concerning types of materials, construction methods, finishes, etc., usually accompanies the set of drawings. The people constructing the building must follow these **specifications** or "specs.," as they are popularly called. The "specs." sometimes allow the builders a choice of materials or methods. Certain materials may not be available. The architect may, therefore, suggest equivalent materials. See Fig. 287.

DESCRIPTION OF MATERIALS

9. PARTITION FRAMING:
Studs: Wood, grade and species #1 hemlock or fir Size and spacing 2x4 16" o.c. Other

10. CEILING FRAMING:
Joists: Wood, grade and species #1 fir or Y.P. Other ; Bridging solid

11. ROOF FRAMING:
Rafters: Wood, grade and species #1 fir or Y.P. ; Roof trusses (see detail): Grade and species

12. ROOFING:
Sheathing: Grade and species #2 fir or Yellow pine size 1"x6" ; type d. & m. ; size "o.c.; solid; spaced ; nailed
Roofing white asphalt shingle ; grade thick ht ; weight or thickness 210# ; size ; fastening nailed
Stain or paint ; Underlay ; number of plies ; surfacing material
Built-up roofing ; number of plies 1602 ; gage or weight 1602 ; gravel stops; snow guards
Flashing: Material copper ; gage or weight ; snow guards

13. GUTTERS AND DOWNSPOUTS:
Gutters: Material ; gage or weight ; size ; shape
Downspouts: Material ; gage or weight ; size ; shape ; number
Downspouts connected to: Storm sewer; sanitary sewer; dry-well; Splash blocks: Material and size

14. LATH AND PLASTER:
Lath walls, ceilings: Material ; weight or thickness ; Plaster: Coats ; finish
Dry-wall walls, ceilings: Material sheetrock ; thickness 3/8 on walls 3/8 on ceilings ; finish ; joint treatment taped

15. DECORATING: *(Paint, wallpaper, etc.)*

Rooms	Wall Finish Material and Application	Ceiling Finish Material and Application
Kitchen	paint (seal & 2 coats)	paint or casein
Bath and all other finished rooms	or papered, except where other finishes are specified	

16. INTERIOR DOORS AND TRIM:
Doors: Type flush ; material birch ; thickness 1 3/8"
Door trim: Type colonial ; material Cl. M. Pine Base: Type 3 member; material Cl. W. pine size 1/4"
Finish: Doors ; trim
Other trim (item, type and location) 2 member 3½ base in garage

17. WINDOWS:
Windows: Type sliding ; make Alwintite division of aluminum
Glass: Grade S.S.B. ; sash weights; balances, type ; head flashing as required
Trim: Type colonial ; material Cl. M. pine Paint ; number coats
Weatherstripping: Type ; material ; Storm sash, number no
Screens: Full half; type top hinge ; number no. ; screen cloth material
Basement windows: Type top hinge ; material M.pine ; screens, number no Storm sash, number no
Special windows plate glass for aluminum sliding doors with the option for
Shutters: Hinged; fixed. Railings Louvers fixed

18. ENTRANCES AND EXTERIOR DETAILS:
Main entrance door: Material Cl. M.P. ; width 3'.0" ; thickness 1-3/4" ; Frame: Material wood ; thickness 1 "
Other entrance doors: Material Cl. W.P. ; width various ; thickness 1-3/4" ; Frame: Material redwood ; thickness 1"
Head flashing as required Weatherstripping: Type interlocking ; saddles metal
Screen doors: Thickness "; number no. ; screen cloth material Storm doors: Thickness "; number no
Combination storm and screen doors: Thickness "; number no.; screen cloth material
Exterior millwork: Grade and species clear redwood Paint ; number coats
1 set of aluminum sliding doors

19. CABINETS AND INTERIOR DETAILS:
Kitchen cabinets, wall units: Material clear white pine ; lineal feet of shelves 27 ; shelf width 11½
Base units: Material clear W. pine ; counter top formica ; edging stainless steel
Back and end splash formica Finish of cabinets ; number coats
Medicine cabinets: Make metal ; model
Other cabinets and built-in furniture vanity in bath - formica top

20. STAIRS:

Stair	Treads		Risers		Strings		Handrail		Balusters	
	Material	Thickness	Material	Thickness	Material	Size	Material	Size	Material	Size
Basement	#1 WP	1-1/8"	#1 WP	3/4"	#1 WP	2x12	W.P.	2x12	#1 WP	1-1/8
Main								2x4		x4 rail
Attic										
Disappearing: Make and model number										

Fig. 287.

CHAPTER TWELVE

STRUCTURAL DRAFTING

Many buildings have frameworks made of steel pieces fastened together. Cranes, radar reflectors or cabinets for electronic equipment, and other similar objects are also composed of steel pieces. Supports for machinery and machine parts themselves have steel shapes that are assembled to form a strong structure. **Structural drawings** show how the construction company must fasten the pieces together.

Many materials other than steel, including aluminum and wood, are used by designers for structural members. However, our discussion shall be limited to steel because it is so widely used.

The structural draftsman must be skilled in both designing and drawing. As a beginner, the structural draftsman may merely draw the simple details or various steel shapes. Or he may show the location of rivet or bolt holes. If he wants to do more than the simplest drawings, he must learn about the individual structural shapes, about rivets and bolts, and how steel shapes may be attached to each other. Buildings and bridges made with steel members require a knowledge of the **strength of materials, reactions, moments, shear tension, compression stresses, welding techniques,** etc. If the draftsman knows and understands these things, he will be able to solve many designing problems on his own or make suggestions to the designer. He will be able to choose a structural shape with the proper size and weight or determine the least number of rivets required in any given situation.

TYPES OF STEEL

Steel is made from iron. When such elements as carbon, silicon, nickel, manganese, molybdenum, or chromium are added, the steel may become stronger or easier to bend or resistant to rust. The steel manufacturer controls the addition of these elements most carefully. The strength of each of the many types of steel has been tested and coded.

The required strength of a structure depends upon such factors as **location, size, type** and, of course, the **weight and size of the material the structure will have to support.** The decision relating to how strong the structure must be is made by the designer. The designer specifies the type or types of steel that will be required, using the proper code number as, for example, ASTM A7. ASTM stands for American Society for Testing Materials. The ASTM determines the strength of each type of steel and codifies it. All the codes can be found in code books published by the ASTM. The draftsman must note on his drawing the code number for each type of steel required for construction.

STEEL SHAPES

Many shapes of steel have been designed for different uses. Table IV lists the name, symbol, size, shape, and appropriate remarks for some of the most common structural shapes. Rolling mills produce these various shapes, called **plain material,** and send them to fabricating shops. The shop personnel cut them to size and drill holes in the proper places for rivets according to the purchaser's order. The fabricating shops also assemble units, such as trusses, or produce special shapes according to the drawings sent to them by the structural designer or draftsman. These are detail shop drawings. Plans, elevations, sections, and details for an entire object or building go to the construction people. The draftsman must put all of the necessary information on all these drawings so that the details can be constructed and the object or building can be erected according to the designer's intentions.

USUAL METHOD OF BILLING AND SKETCHING STRUCTURAL STEEL SHAPES ON SHOP DRAWINGS.

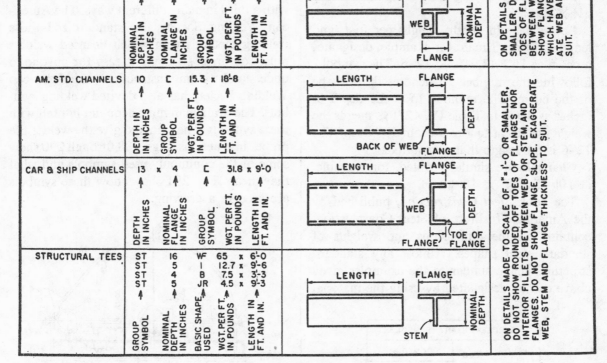

GROUP	EXAMPLE OF BILLING ON DETAIL BILLING	CONVENTIONAL WAY OF SHOWING ON DETAIL DRAWINGS AND IDENTIFICATION OF MAJOR COMPONENT PARTS	REMARKS
EQUAL LEG ANGLES	L 3 1/2 x 3 1/2 x 1/4 x 5'-6 — GROUP SYMBOL, LEG WIDTH IN INCHES, LEG WIDTH IN INCHES, THICKNESS IN INCHES, LENGTH IN FT. AND IN.	LENGTH, LEG, LEG, TEE, FILLET, THICKNESS, TEE	ON DETAILS MADE TO SCALE OF 1"=1'-0 OR SMALLER, DO NOT SHOW ROUNDED OFF TOES OF ANGLES OR INTERIOR FILLET BETWEEN LEGS. BILL LONG LEG OF UNEQUAL LEG ANGLES FIRST. EXAGGERATE LEG THICKNESS TO SUIT.
UNEQUAL LEG ANGLES	L 6 x 4 x 3/8 x 10'-3 — GROUP SYMBOL, LONG LEG IN INCHES, SHORT LEG IN INCHES, THICKNESS IN INCHES, LENGTH IN FT. AND IN.	LENGTH, SHORT LEG, LONG LEG, THICKNESS	
AM. STD. I-BEAMS 15 I 42.9 x 16'-3 1/2 MISC. LIGHT BEAMS 12 B 19 x 8'-2 1/4 STD. MILL BEAMS 8 M 17 x 7'-10 JUNIOR BEAMS 12 JR 11.8 x 9'-3 JOISTS 8 J 10 x 7'-0 WIDE FLANGE BEAMS 27 WF 94 x 26'-10 BEARING PILES 12 BP 74 x 20'-3 — NOMINAL DEPTH, GROUP SYMBOL, WGT. PER FT. IN POUNDS, LENGTH IN FT. & INCHES	LENGTH, FLANGE, WEB, NOMINAL DEPTH, FLANGE	ON DETAILS MADE TO SCALE OF 1"=1'-0 OR SMALLER, DO NOT SHOW ROUNDED OFF TOES OF FLANGES OR INTERIOR FILLETS BETWEEN WEB AND FLANGES. DO NOT SHOW FLANGE SLOPE FOR THOSE BEAMS WHICH HAVE SLOPING FLANGES. EXAGGERATE WEB AND FLANGE THICKNESS TO SUIT.
MISC. LIGHT COLS.	8 x 8 M 34.3 x 10'-11 — NOMINAL DEPTH IN INCHES, NOMINAL FLANGE IN INCHES, GROUP SYMBOL, WGT. PER FT. IN POUNDS, LENGTH IN FT. AND IN.	LENGTH, FLANGE, WEB, NOMINAL DEPTH, FLANGE	
AM. STD. CHANNELS	10 C 15.3 x 18'-8 — DEPTH IN INCHES, GROUP SYMBOL, WGT. PER FT. IN POUNDS, LENGTH IN FT. AND IN.	LENGTH, FLANGE, DEPTH, BACK OF WEB, FLANGE	ON DETAILS MADE TO SCALE OF 1"=1'-0 OR SMALLER, DO NOT SHOW ROUNDED OFF TOES OF FLANGES NOR INTERIOR FILLETS BETWEEN WEB, OR STEM, AND FLANGES. DO NOT SHOW FLANGE SLOPE. EXAGGERATE WEB, STEM AND FLANGE THICKNESS TO SUIT.
CAR & SHIP CHANNELS	13 x 4 C 31.8 x 9'-0 — DEPTH IN INCHES, NOMINAL FLANGE IN INCHES, GROUP SYMBOL, WGT. PER FT. IN POUNDS, LENGTH IN FT. AND IN.	LENGTH, FLANGE, WEB, DEPTH, TOE OF FLANGE, FLANGE	
STRUCTURAL TEES	ST 16 WF 65 x 6'-0 ST 5 I 12.7 x 9'-6 ST 4 B 7.5 x 9'-3-3 ST 5 JR 4.5 x 9'-3-3 — GROUP SYMBOL, NOMINAL DEPTH IN INCHES, BASIC SHAPE USED, WGT. PER FT. IN POUNDS, LENGTH IN FT. AND IN.	LENGTH, FLANGE, NOMINAL DEPTH, STEM	

Table IV.

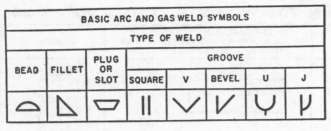

Fig. 288.

Fig. 289.

DESIGNATING STEEL SHAPES

Standard methods of indicating steel pieces have been devised. The symbol, found in Table IV, is followed by dimensions. The dimensions vary according to the shape. An angle, for example, has two legs of a certain thickness. The legs may not always be of equal length. Therefore, the length of each leg and the thickness of both legs are given. For example, an angle whose leg lengths are equal would be indicated by the notation $\angle$ 3×3×¼. An angle with unequal leg lengths might be designated as $\angle$ 3½×2½×¼.

Beams vary in depth, weight per foot, and length. The draftsman, for example, designates a beam as 15 I 42.9 ⚡ × 12′-6. The symbol *I* following the number *15* denotes the *I* shape of the beam. The number *15* is the **depth in inches** as noted in Table IV. 42.9 ⚡ means the **weight per lineal foot** of a beam 15″ deep. The 12′-6 is the **length** that is required.

Channels are similarly noted, for example, as 10[15.3 ⚡ × 18.8.

The book *Steel Construction*, published by the American Institute of Steel Construction, contains all the dimensions and weights of standard steel shapes. Almost any standard structural shape similar to the examples given above can be designated by using the material

contained there. All the dimensions necessary for each shape must be included in the drawing.

METHODS OF FASTENING STEEL SHAPES

Welding. Welding is the process of first melting two pieces of steel to be joined at a desired spot and then fusing them with a third material. Welding provides the strongest type of fastening but is very costly.

Two pieces of steel can be melted and fused with a third in many different ways. The American Welding Society has standardized those welding methods with which licensed welders must conform. To make it easy for anyone to understand welding instructions, the American Welding Society has also devised welding symbols. The draftsman must acquaint himself with these symbols if he wants to give the welder the proper instructions. Figs. 288 through 290 show some of the commonly used symbols and what they mean. Fig. 291 shows how these symbols are used in a drawing.

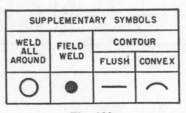

Fig. 290.

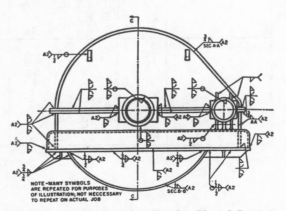

Fig. 291. Typical Drawing Showing Use of Symbols.

Riveting. Two pieces of steel can be joined by drilling matching holes in each piece and inserting a rivet through the holes. The rivet may be a solid piece of steel or a tube. It may have a rounded or flat head on one end. The other end sticks out and is pounded down. Figs. 292 through 294 illustrate the various ways of pounding.

Standards, which have code numbers, have been developed for both sizes and materials for all types of rivets. The draftsman must indicate on the drawing the sizes and code numbers of the rivets to be used. Sometimes they are listed on a separate sheet. Fig. 295 illustrates some of the symbols used for rivets, and Fig. 296 shows how rivets are dimensioned.

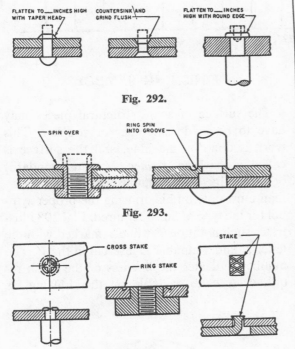

Fig. 292.

Fig. 293.

Fig. 294.

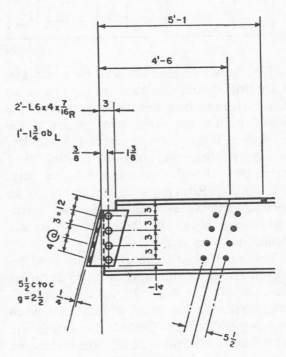

Fig. 296.

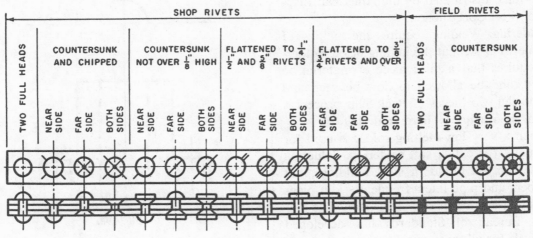

Fig. 295.

USUAL GAGES FOR ANGLES, INCHES

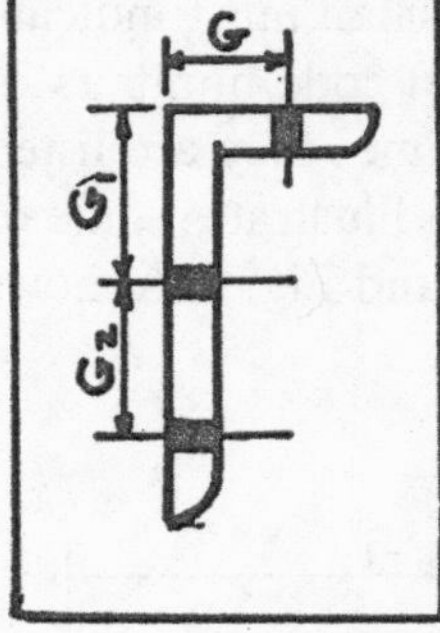

LEG	8	7	6	5	4	3½	3	2½	2	1¾	1½	1⅜	1¼	1
G	4½	4	3½	3	2½	2	1¾	1⅜	1⅛	1	⅞	⅞	¾	⅝
G₁	3	2½	2¼	2										
G₂	3	3	2½	1¾										

Fig. 297.

Fig. 297 shows gage distances for rivets. The holes drilled in the **flange** of the steel piece for the rivets must be a certain distance from the **web.** The distance varies with the size of the steel piece. If two pieces of steel of different sizes require riveting, the holes for riveting must meet, even though their gage distances vary. See Fig. 297. Rivet holes are usually 1/16 of an inch larger than the diameter of the rivet. Standard spacing of the rivet holes for various size beams has been devised.

Bolting. High-strength steel bolts have replaced rivets, in many instances, in fastening steel shapes. The technique for using bolts in construction is the same as that used when working with rivets. The bolt dimensions will be similar to Regular Semifinished Hexagon bolts. For complete specifications for High Strength Steel Bolts, see the pamphlet *Assembly of Structural Joints Using High Strength Steel Bolts,* distributed by the American Institute of Steel Construction.

Soldering. Welding requires the melting of two pieces of steel at the joining place. **Soldering** requires that a third piece of material be melted, and be allowed to flow between and around the two pieces of steel. This process is known as **brazing** or **hard soldering.**

When a very low-melting solder is used for fastening, it is called **soft soldering.** Soldering will seldom be found in steel buildings, but plumbers and pipe fitters frequently join pipes and fittings with soft or hard solder. The National Bureau of Standards has developed standards for all types of soldering.

FINISHING SYMBOLS

The surface of some structural pieces may have to be made smoother or rougher. This work is done by machine, and the surface is called a **machine surface.** Various standards or finishes have been established, and the draftsman must note on his drawing the proper symbol for the type of finish required. Fig. 298 illustrates such notations. A line is marked with the letter *V* and a number is inserted in the *V*. The number indicates the **fineness** of the finish required. A code or table on the drawing, as

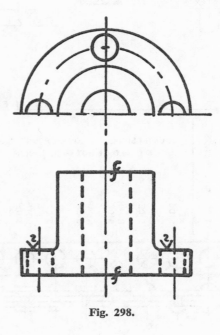

Fig. 298.

shown in Fig. 291, should explain the exact meaning of the numbers. Each company may have its own finishing standards. Finishing symbols are found on drawings for steel objects or steel machines, but they are seldom found on drawings for steel buildings.

MAKING A STRUCTURAL DRAWING

Structural drafting includes two categories. In one, the shop detail drawing, the draftsman draws the details from the designer's sketch. This drawing goes to the fabricating shop, where the small assemblies are made, the shapes cut to the desired lengths, and the holes for rivets or bolts drilled. These constructed parts are then sent to the construction site, where the entire structure is erected according to the second category of drawings. This category includes plans, elevations, sections, special details of the entire building, and large areas of the building which were sketched by the designer and drawn to scale by the draftsman.

SHOP DETAIL DRAWINGS

Fabricating shop detail drawings may include any of the following:
1. Simple Square-Framed Beams
2. Special Framed-Beam Connections
3. Seated Connections
4. Columns
5. Skewed Connections
6. Welded Beams
7. Truss and Bracings
8. Roof Purlins
9. Eave Struts
10. Girt Framing
11. Sash
12. Lateral Bracing
13. Crane Runway
14. Lintel Beam
15. Laced Box Strut
16. Crane Column
17. Welded Truss and Bracing
18. Beam and Girder Bridge
19. Truss Bridge

Simple Square-Framed Beam-Connection Detail Drawing. Fig. 299 shows a designer's plan view of a simple square-framed beam connection, and Fig. 300 shows a detail drawing of part of that plan. The **framed beam connection** means that a short angle joins the two beams. The plan view includes the north arrow, the distance—20 feet—between the center lines of the supporting beams, the size and weight of each beam, and the elevation at the top of each beam from the ground, or grade, or whatever horizontal line is used as a base from which elevations are measured. The plus sign means that the elevation is above the horizontal base line.

The detailer, in Fig. 300, must show how the *24WF* and *14WF* connect. The book *Steel Con-*

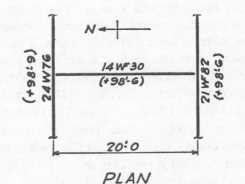

PLAN

Elev. top of steel shown thus (+98'-6)

Fig. 299.

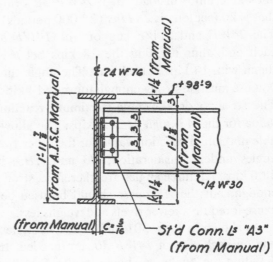

Fig. 300.

struction, known as The Manual, shows the depth of the *24WF* as 1'-11⅞. The Manual is vital to the structural draftsman. When assembling the three beams, as shown in the plan view, Fig. 299, the construction crew puts the two supporting beams, the *24WF* and the *21WF*, in first. The *14WF* must be somewhat shorter than 20 feet in order to get it between the two supporting beams. Standard shortening distances will be found in the manual. The web thickness of the supporting beams will determine how much shorter the *14WF* will have to be. The manual shows that the **shortening distance,** called the **c distance,** for the *24WF* is 5/16 of an inch.

The detailer will also determine the size of the angle which connects the *24WF* and the *14WF*. The angle will have to be as strong as the *14WF*. The manual shows a standard angle *A3* as suitable for a *14WF 30* beam. The *A3* standard angle also meets the designer's specifications for ⅞-inch rivets. However, the manual also shows that *angles H3* and *HH3* are also suitable for the *14WF 30* beam. The latter two larger angles require more rivets and are heavier than the *A3* angle. If the *A3* angle is strong enough, that is the one that should be used. It will be less expensive, require less riveting, and will reduce the weight of the entire structure.

How does the detailer determine which angle to use? The manual shows that the maximum allowable uniform load for a *14WF 30* beam that is 20 feet long is *28 kips* (28,000 pounds). The *24WF* and *21WF* support the *14WF* at each end, thus dividing the 28 kips between them, with 14 kips at each end. The angle that you use must be strong enough for only 14 kips. The **strength** or **allowable maximum reaction value** for the *A3 angle* is 28.4 kips. The allowable maximum reaction value for the other two angles under consideration, *H3* and *HH3*, is 48.6 kips. All three angles are, therefore, strong enough, but the *A3* angle should be used because it requires fewer rivets and weighs less.

As shown in Figs. 301 and 302, two angles, one on each side of *14WF 30*, are needed. In detailing the angle, as shown in Fig. 303, the spacing of the rivet holes, both vertically and horizontally, is shown. Angle *A3* is 8½ inches long. The vertical distance between rivets, called the pitch, is 3 inches. Since there are two distances, each 3 inches, between rivets, there still remains 2½ inches from the over-all length of 8½ inches. Therefore, center the rivet holes by leaving 1¼ inches at each end.

For the horizontal location of the rivet holes, the distance from the gage line is used. Rivets appear in a straight, vertical line called the **gage line,** shown in Fig. 303. Standard gages, as shown in Fig. 297, have been devised for various angle-leg sizes. However, fabricators depart from these standards. The structural

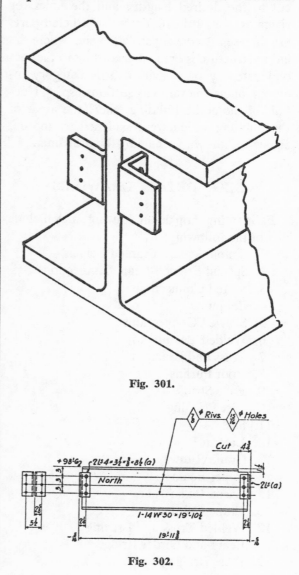

Fig. 301.

Fig. 302.

draftsman will probably use his company's standards and refer to the standard gages, shown in Fig. 297, only when company standards do not adequately meet his needs.

In Fig. 303, the 5½-inch dimension is called the **spread.** The open holes are for the web rivets, and the smaller black dots are for the rivets that go into the supporting beam, the *24WF 76.* The gage is found on the 4-inch leg by the following method:

$$\frac{1}{2} \ (5\frac{1}{2}'' - \frac{1}{4}'') = 2\frac{5}{8}''$$
$$G = 2\frac{5}{8}''$$

Place the angles on the *14WF beam* so that the center line of the top rivet hole measures 3 inches from the top of the *14WF beam,* as shown in Fig. 300. The manual gives 2¼ inches rather than 3 inches. However, wherever practicable, 3 inches should be used for all beams to make the work simpler.

When the *14WF beam,* with the *A3 angles* riveted to it, is put against the *24WF,* the draftsman must make sure that the corner of the *14WF* beam will not bump into the inside corner of the *24WF.* The distance between the top of the *24WF* and the top of the *14WF* is 3 inches. The manual gives the **fillet, or k distance,** as 1¼ inches. This means that the corner of the *14WF* need not be cut because it will not hit the *24WF.*

At the south end of the plan view, Fig. 304, the connection of the *14WF 30* to the *21WF 82* beam must be shown. Use the same procedure just described for the north-end construction. The important difference between the north- and south-end constructions concerns the top elevations of the beams. In the south end the *14WF* and *21WF* appear at the same elevation. The top corner of the *14WF* must, therefore, be cut away to keep it from hitting the inside corner of the *21WF.* See Fig. 304. This cutaway is called a **cope, block,** or **cut.** The distance *Q1* equals the *k distance* for *21WF 82.* Distance *Q2* is found by adding ½ inch to the *a distance* for *21WF 82,* which can be found in the manual. The distance *Q2* equals 4¾ inches.

For the complete detail drawing, see Fig. 302. The *A90* in the title is an identification for shipping purposes.

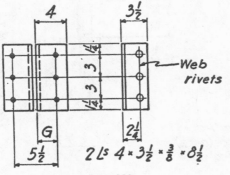

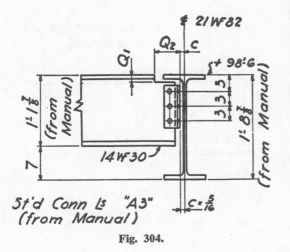

Fig. 303.

Fig. 304.

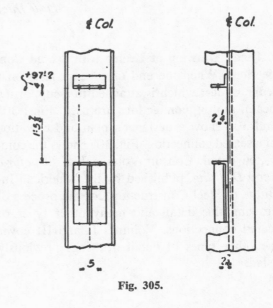

Fig. 305.

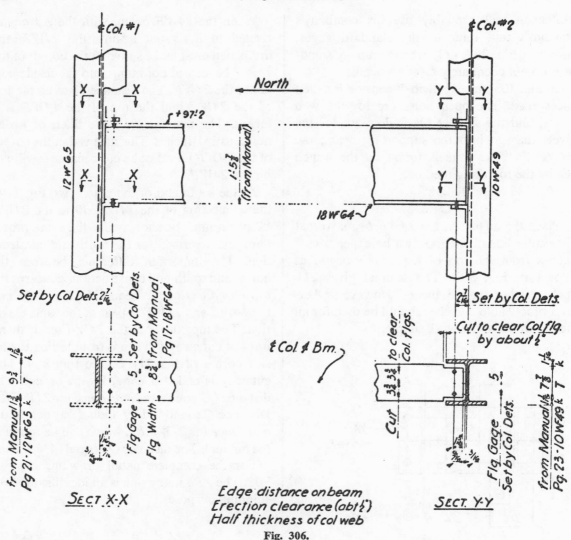

Edge distance on beam
Erection clearance (abt ½)
Half thickness of col web

Fig. 306.

Detail Drawing of Beams with Seated Connections. When the end of a beam rests on a **ledge** or **seat** which is attached to a supporting beam, **seating connections** are used. Figs. 305 and 306 show a designer's plan and elevation of a seated connection. Fig. 307 shows the completed detail. Consult Volume I of *Structural Shop Drafting,* published by the American Institute of Steel Construction, for the process of drawing this detail and many other types of seated connections. Volumes II and III cover the other types of detail drawings previously listed.

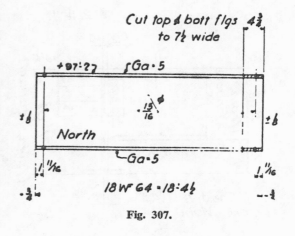

Fig. 307.

PLANS, ELEVATIONS, SECTIONS, AND LARGE DETAILS FOR THE ENTIRE BUILDING

The structural draftsman may not draw the type of details previously discussed. He may be involved in plans, elevations, sections, and large details dealing with entire floors or the whole building. The construction crew, not the fabricating shop, uses these drawings.

Plans. Plans for steel members of industrial buildings are drawn for each floor and for any other levels which might have special steelwork. The ground floor is called the first floor. First-floor plans appear on a drawing board called the **column schedule.** The column schedule shows the **location** and **numbering** of all vertical columns in the entire building. In a diagrammatic elevation, also shown on the column schedule, the actual heights of these vertical columns up through the entire building are given. Fig. 308 is a typical column-schedule drawing. As Fig. 308 shows, the second-floor plans show the steel pieces on and at the second floor, and the third-floor plans show the steel pieces on and at the third floor, etc. **The elevations of the top of the steel members appear in the title of each plan.** If a plan view of a second floor contains pieces a little below or above the second floor, they are included in the drawing but their elevations are given separately. If special steel pieces are necessary for supporting machinery or other pieces of equipment, they are included in the plan of the floor on which they will be placed. Vertical columns should also be shown in all plan views. An arrow pointing north should be put on all plan views but not on elevations or sections.

Elevations and Sections. Elevations and sections appear on separate drawings. See Fig. 309. North, south, east, and west elevations show the steel from each side of the building. The north elevation, for example, shows the north side of the building. The observer looks south.

Sections, of course, show parts of the building which have special steelwork. All plans, elevations, and sections show only the **type, size,** and **location** of steel pieces. They do not show

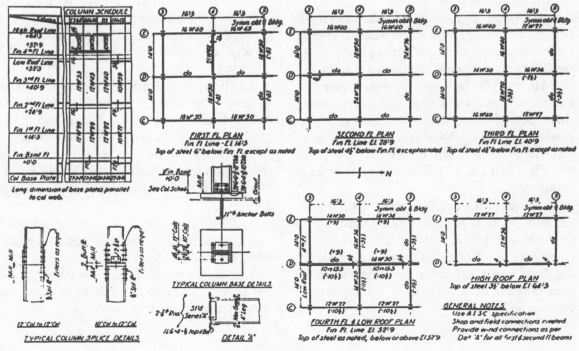

Fig. 308.

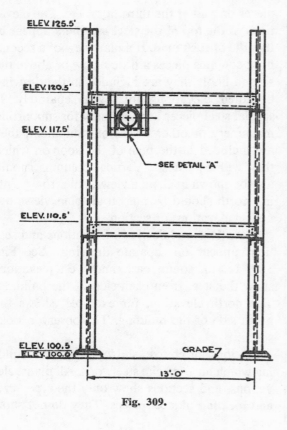

Fig. 309.

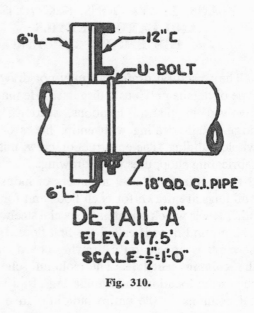

DETAIL "A"
ELEV. 117.5'
SCALE—½":1'0"

Fig. 310.

the methods of fastening, because such details are left for detail drawings.

Large Detail Drawings. If all the steel members of the building are to be welded, riveted, or bolted, a note to that effect is put on all the steel drawings. If some members are to be fastened in a special way, details of these members might be drawn on a separate detail-drawing sheet. Some parts of the structure often require **reinforcing,** such as **knee bracing** or **wind bracing.** The draftsman will draw these parts in detail. Figs. 310 and 311 illustrate some types of details.

The scale of the detail depends upon the object to be drawn. A scale larger than the plan or elevation might illustrate details more easily. In plan views of large buildings, one scale for the depth and another for the length might be used in order to fit the whole building on one drawing sheet.

Steel pipes that slope must be labeled as sloping, and the slope, in inches per foot, must be given. A triangle is sometimes used to illustrate the slope. For example, one leg of the triangle might be labelled 3 inches and another 12 inches, indicating a slope of 3 inches per foot.

The structural draftsman follows the general rules of dimensioning discussed in Chapter Three and illustrated in Figs. 296 and 311. For general rules, the following points should be observed:

1. Dimensions are given in feet and inches. If the dimension is less than a foot, simply write the inch dimension as 7″, not as 0′-7″. Omit the inch mark when giving feet and inches.

2. When dimensions of 1/32″ or 1/64″ are used with foot dimensions, they are rounded to the nearest ⅛″ unless such dimensions are critical.

3. Dimensions are given from the center line of one steel beam to another, and to the backs of angles and channels.

4. Number the columns in one direction, for example, north, and letter them in another direction, such as west, as shown in Fig. 308.

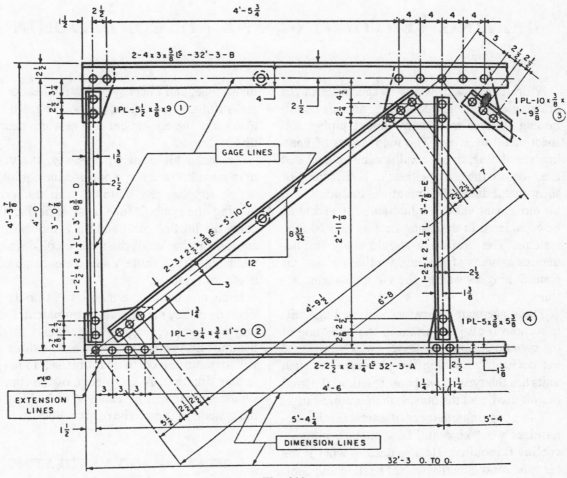

Fig. 311.

HEATING, VENTILATING, AND AIR CONDITIONING

A draftsman who wants to specialize in the drawing of heating, ventilating, and air-conditioning systems must know about piping and sheet-metal drawing. The many ways of heating, ventilating, or air-conditioning an enclosed area require the use of either, or both, piping or sheet-metal ducts. Temperature, humidity, and air movement affect the human body and must be considered in designing heating and cooling systems. The draftsman should study the numerous systems of heating, ventilating, and air conditioning as well as the equipment and fittings.

The placement of heating and cooling units in a room is also important in the designing of the system. Among the factors which influence the decision governing the placement of such units are the type of wall construction, the kind of unit used, and the outside temperature range.

The designing engineer of each building determines what kind and how much heating or cooling is required. He also decides what is the best and most effective way of heating and cooling the building and how and where the heating and cooling systems will be distributed.

The engineer must first find out how much heat escapes from a building. The type of building construction, the velocity of prevailing winds, and the degree of outside temperature affect the loss of heat in a building. When the engineer determines the amount of heat that is lost, he will know how much heat will be required to replace it.

Heat is measured in terms of *BTU* or *British Thermal Units*. One *BTU* equals the amount of heat necessary to raise the temperature of one pound of water to one degree Fahrenheit. *MBH* stands for *1000 BTU per hour*.

The draftsman must draw, in its proper location, the equipment which makes a building warm or cold. He must also draw the equipment, pipe, and ducts by which the heat or cold is distributed throughout the building. He must also show the equipment that services each individual area.

To make his drawing properly, the draftsman must know the symbols for the equipment which supplies the building with the proper heating, the symbols for the pipes or ducts that carry the heat or cold, and the symbols for the equipment that sends the heat or cold into each particular area. Table V illustrates some of the standard symbols.

Some companies use their own symbols, particularly if they manufacture equipment. Others may use simple, single-line drawings to show heating and ventilating systems. The draftsman will, of course, have to adapt himself to his particular situation. A knowledge of the standard symbols contained in Table V and of some of the systems in use will be of great help.

HEATING AND VENTILATING SYSTEMS

Gravity and Forced Warm Air Systems. Air comes in from the outside or comes back after being used in a structure, enters a furnace or heater in the basement, goes through pipes or ducts in the building and out into the room. This is called the **Gravity Warm Air System.** When fans force the air into the heaters, it is known as the **Forced Warm Air System.**

The heaters may be any of the following:
1. **Boilers** made of steel or cast iron
2. **Furnaces**
   a. For the Gravity System
   b. For the Forced Air System
3. **Space Heaters**
   a. Solid Fuel, heated by fire
   b. Oil Heaters
   c. Gas Heaters

PIPING		TOP & BOTTOM REGISTER OR GRILLE	T & BR - (20x12-700 CFM) EA. / T & BG - (20x12-700 CFM) EA.
REFRIGERANT DISCHARGE	—— RD ——	CEILING OUTLET OR GRILLE	CO - 20 x 12 - 700 CFM / CG - 20 x 12 - 700 CFM
REFRIGERANT SUCTION	— — RS — —	LOUVER OPENING	L → 20 x 12 - 700 CFM
CONDENSER WATER SUPPLY	—— C ——	ADJUSTABLE PLAQUE	P - 20x12 700 CFM / P - 20 φ 700 CFM
CONDENSER WATER RETURN	— — CR — —		
CIRCULATING CHILLED OR HOT WATER SUPPLY	—— CS/HS ——	VOLUME DAMPER	PLAN / ELEV.
CIRCULATING CHILLED OR HOT WATER RETURN	—— CR/HR ——		
MAKE - UP WATER	—— · — MW — · ——	DEFLECTING OR SPLITTER DAMPER	
HUMIDIFICATION LINE	—— · —— H —— · ——		
DRAIN LINE	—— D ——	DEFLECTING OR SPLITTER DAMPER - UP	
BRINE SUPPLY	—— B ——	DEFLECTING OR SPLITTER DAMPER - DOWN	
BRINE RETURN	— — BR — —		
DUCTWORK		ADJUSTABLE BLANK OFF	TR / 20 x 12
DUCT (FIGURES ARE SIZE OF DUCT IN VIEW SHOWN)	12 x 20	TURNING VANES	
DIRECTION OF AIR FLOW	→		
INCLINED DROP IN DIRECTION OF AIR FLOW	⫴→ DN. ⫴		
INCLINED RISE IN DIRECTION OF AIR FLOW	UP ⫴→ ⫴	MULTIBLADE DAMPERS / M = DENOTES AUTOMATIC OPERATION	
SUPPLY DUCT SECTION	S ←12 x 20		
EXHAUST DUCT SECTION	E ←12 x 20	MULTIBLADE DAMPERS (THROTTLING TYPE) / M = DENOTES AUTOMATIC OPERATION	
RECIRCULATING DUCT SECTION	R ←12 x 20		
FRESH AIR DUCT SECTION	F A ←12 x 20		
REGISTER	"R"	FLEXIBLE CONNECTIONS	
GRILLE	"G"		
SUPPLY OUTLET	⫴→		
EXHAUST INLET	⫴←//	INTAKE LOUVERS & SCREEN	
TOP REGISTER OR GRILLE	TR - 20 x 12 - 700 CFM / TG - 20 x 12 - 700 CFM		
CENTER REGISTER OR GRILLE	CR - 20 x 12 - 700 CFM / CG - 20 x 12 - 700 CFM		
BOTTOM REGISTER OR GRILLE	BR - 20 x 12 - 700 CFM / BG - 20 x 12 - 700 CFM		

Table V.

After the air leaves the heater, pipes or ducts, called **leaders,** in the basement take it to vertical pipes, called **stacks.** Stacks are usually found inside partitions of buildings. The stacks take the air to the various areas in the building. Each area to be heated has openings, called **register boxes,** in the floor or on the side wall near the floor. The register boxes contain registers which distribute the flow of warm air into the room as it comes from the stacks.

Various fittings are necessary to put this kind of system together properly. Fig. 312 illustrates some commonly used Warm Air System fittings.

Steam-Heating Systems. Steam as well as warm air can be used to heat a building. Steam is made by heating cold water in a boiler until it boils and then turns to steam in a converter. Pumps force the steam into pipes which are located throughout the building. In each room or area requiring heat, a radiator, finned-tube unit, baseboard unit, or convector is attached to one of the steam-carrying pipes.

Radiators are generally made of cast iron, and may be in the form of a column or riser, a large tube, or a wall type. Fig. 313 shows **one-pipe radiator connections,** Fig. 314 shows **two-pipe top-and-bottom opposite-end radiator connections,** and Fig. 315 shows **two-pipe connections to a radiator hung on a wall.** A **finned-tube unit** is a metal tube with many metal fins attached to it, as shown in Fig. 316. A **baseboard unit,** shown in Fig. 317, is found along the bottom of a wall, with its front part exposed to the room. A **convector,** shown in Fig. 318, works by combining steam and air.

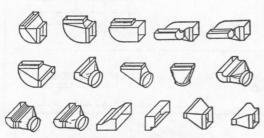

Fig. 312. Warm Air System Fittings.

Reprinted by permission from HEATING VENTILATING AIR CONDITIONING GUIDE, 1958, Chapter 19, p. 503.

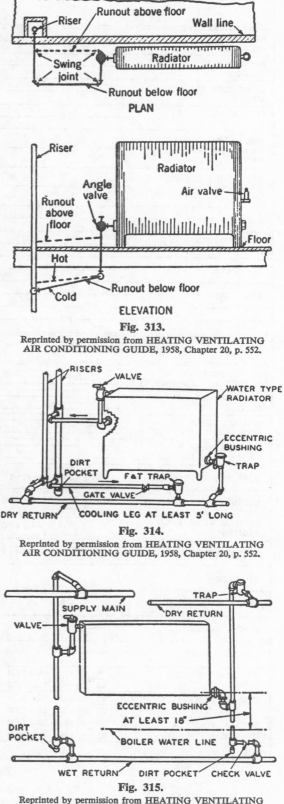

PLAN

ELEVATION

Fig. 313.

Reprinted by permission from HEATING VENTILATING AIR CONDITIONING GUIDE, 1958, Chapter 20, p. 552.

Fig. 314.

Reprinted by permission from HEATING VENTILATING AIR CONDITIONING GUIDE, 1958, Chapter 20, p. 552.

Fig. 315.

Reprinted by permission from HEATING VENTILATING AIR CONDITIONING GUIDE, 1958, Chapter 20, p. 552.

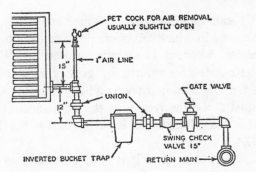

Fig. 316.
Reprinted by permission from HEATING VENTILATING
AIR CONDITIONING GUIDE, 1958, Chapter 20, p. 553.

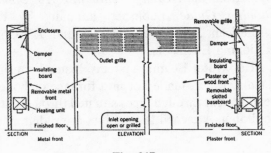

Fig. 317.
Reprinted by permission from HEATING VENTILATING
AIR CONDITIONING GUIDE, 1958, Chapter 22, p. 599.

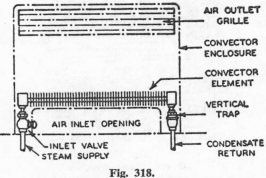

Fig. 318.
Reprinted by permission from HEATING VENTILATING
AIR CONDITIONING GUIDE, 1958, Chapter 20, p. 553.

A heating engineer determines which of these steam-heat distributors to use. On each distributor there is a valve with which to regulate the amount of steam going into the distributor from the pipes.

Although the steam-heating process just described is a basic one, there are several different names for it. The systems vary and these variations determine the name of the particular system. For instance, when only one pipe serves the steam to the radiator and also carries back the condensate from the radiator, it is called a **one-pipe system,** as shown in Fig. 319. Condensate forms in the pipe from some of the steam that cools and turns to hot water. A **two-pipe** system has one pipe to serve the steam and another pipe to carry back the condensate. See Fig. 320.

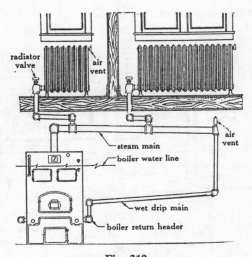

Fig. 319.
Courtesy Dunham-Bush, Inc.

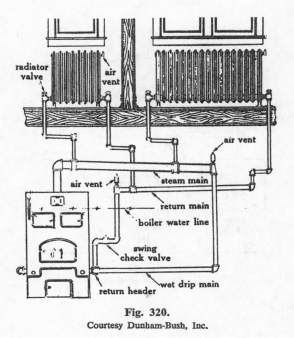

Fig. 320.
Courtesy Dunham-Bush, Inc.

Instead of pipe differences, steam systems may vary in pressure and vacuum conditions. A **low-pressure system** works from zero to 15 pounds per square inch (psi) as shown in Fig. 321. A **high-pressure system** works above 15 pounds per square inch (psi), shown in Fig. 322. A **vapor system,** shown in Fig. 321, works under vacuum and low pressure without the use of a vacuum pump. A **vacuum system** works under vacuum and low pressure with a vacuum pump, as shown in Fig. 323.

Steam-heating systems are named for the manner in which the condensate returns to the boiler, where it is remade into steam. When the condensate returns by gravity, it is called a **Gravity Return system.** When a trap or pump must force the return of the condensate, it is called a **Mechanical Return system.** See Fig. 324.

There are variations and additions to the types of steam-heating systems described above. The draftsman can obtain additional information from the *Heating Ventilating Air Conditioning Guide,* published annually by the American Society of Heating and Air-Conditioning Engineers.

Hot-Water Heating Systems. Instead of using steam, water, made hot in a furnace or boiler, can be sent through pipes situated throughout a building. Attached to the pipes are radiators or

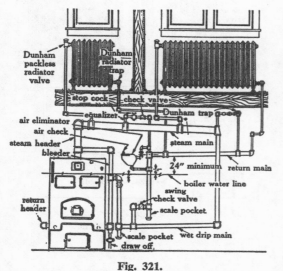

Fig. 321.
Courtesy Dunham-Bush, Inc.

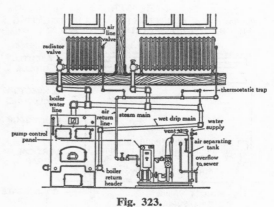

Fig. 323.
Courtesy Dunham-Bush, Inc.

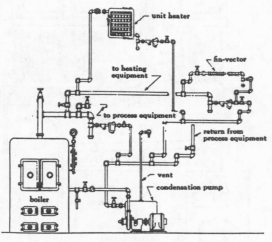

Fig. 322.
Courtesy Dunham-Bush, Inc.

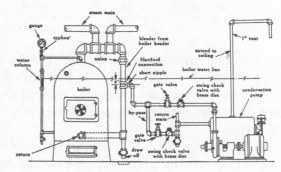

Fig. 324.
Courtesy Dunham-Bush, Inc.

other heat-distributing units located where heat is required.

If the hot water flows due to the difference in weight between the supply and return amounts of water, the system is called a **Gravity system.** The Gravity system is seldom used commercially, but it is used in homes. If the hot water is forced to flow because of a pump, the method is known as a **Forced system.** A Forced system is divided into two temperature systems, a **Low Temperature system** and a **High Temperature system.** When a Forced system operates at a temperature of 250° F or less, it is operating on a Low Temperature system; when it operates at a temperature above 250° F, it is operating on a High Temperature system. Both temperature systems are capable of distributing hot water through three different piping arrangements. When one pipe supplies and returns the hot water, the method of heating is called a **One Pipe system,** as illustrated in Fig. 325. When one pipe supplies the hot water and another carries it away, the method of heating is known as a **Two Pipe system,** as shown in Fig. 326. A third system, the **Series Loop system,** shown in Fig. 327, has one or more loops or circuits. The same hot water circulates in succession through each heating unit or each circuit.

Panel Heating. The **Panel Heating system** is based upon the use of warm water in pipes or tubes, or warm air in ducts or electrical elements. The equipment may be imbedded in ceilings, walls, or floors to heat large areas at temperatures between 80° and 125° F. Fig. 328 shows a typical example of a panel-layout design.

The pipes are 1 inch, ¾ of an inch, or ½ inch in diameter. The tubes are ⅜ of an inch, ⅝ of an inch, or ⅞ of an inch in diameter. The pipes or tubes may be formed in coils and imbedded in concrete and plaster. If this is desired, only the welded-type fittings should be used. Bends should be used instead of elbows. Nonferrous coils and pipes may be soldered.

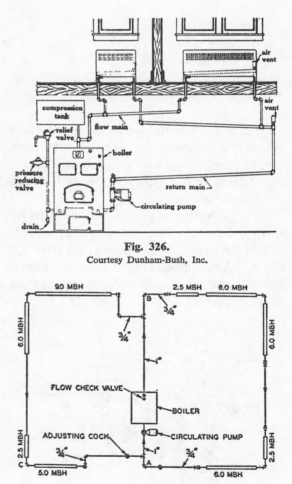

Fig. 326.
Courtesy Dunham-Bush, Inc.

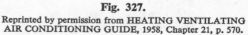

Fig. 327.
Reprinted by permission from HEATING VENTILATING
AIR CONDITIONING GUIDE, 1958, Chapter 21, p. 570.

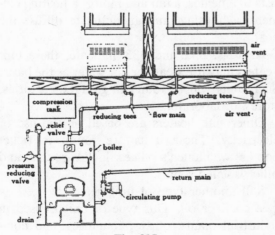

Fig. 325.
Courtesy Dunham-Bush, Inc.

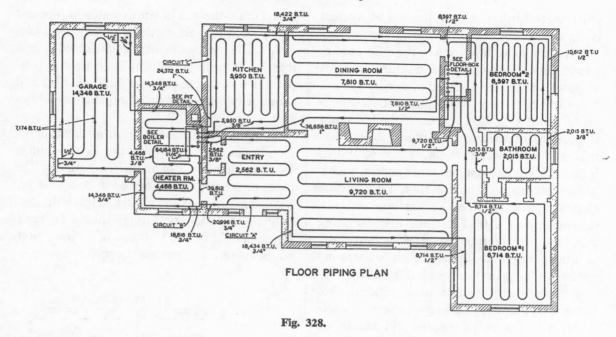

FLOOR PIPING PLAN

Fig. 328.

Pipes in ceilings and walls may be imbedded in concrete or plaster above or below the metal lath, or the pipe may be in prefabricated panels. Fig. 329 illustrates typical floor- and ceiling-panel installations.

Pipes in floors are at least 1½ to 4 inches below the floor, with a waterproofing layer to protect the insulation and piping. See Figs. 329 and 330. Underground panel heating is used to melt snow on roads, sidewalks, ramps, and runways.

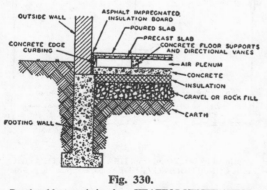

Fig. 330.

Reprinted by permission from HEATING VENTILATING AIR CONDITIONING GUIDE, 1958, Chapter 23, p. 608.

UNIT HEATERS

The engineer usually designs the heating or ventilating system. If his problem is simple or one that he often meets, he saves himself much time and money by buying a commercially made system that satisfies his design problems. These commercially made systems are called units. A **unit heater,** as shown in Fig. 331, consists of a heater, a fan and motor, a heating element, a housing, and an outlet to diffuse the heat.

As Figs. 332 and 333 indicate, these unit heaters may use steam or hot water as the basis. Such units will be found in garages, factories, stores, or laboratories.

Other unit heaters are based on the use of electricity. These units are found in ticket booths, watchmen's offices, locker rooms, and other isolated rooms.

Still another type of unit heater is fired by gas, coal, or oil. This type of unit is used in industrial plants, offices, foundries, and commercial buildings.

The draftsman will have to draw these units. His drawing must include locations, necessary notes, and other pertinent data.

TYPICAL FLOOR AND CEILING PANEL INSTALLATIONS

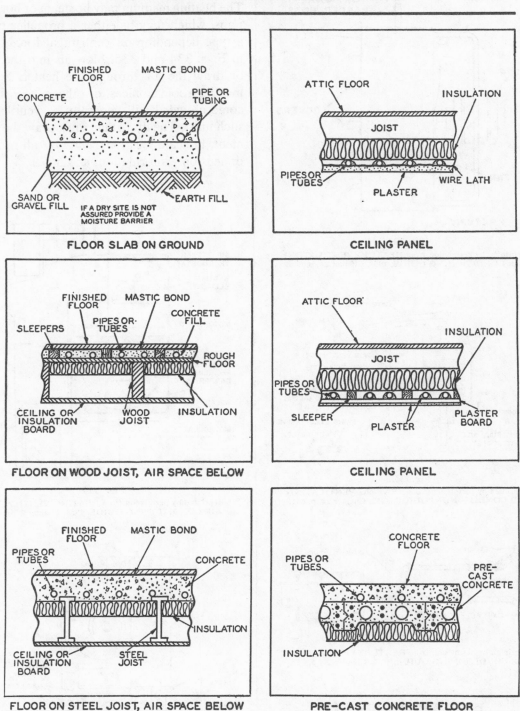

Fig. 329.

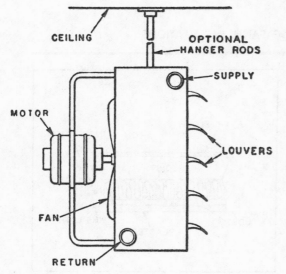

Fig. 331.
Reprinted by permission from HEATING VENTILATING
AIR CONDITIONING GUIDE, 1958, Chapter 24, p. 646.

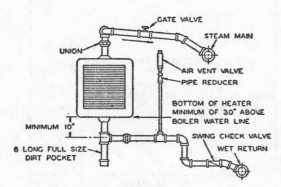

Fig. 332.
Reprinted by permission from HEATING VENTILATING
AIR CONDITIONING GUIDE, 1958, Chapter 24, p. 651.

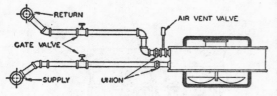
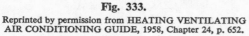

Fig. 333.
Reprinted by permission from HEATING VENTILATING
AIR CONDITIONING GUIDE, 1958, Chapter 24, p. 652.

UNIT VENTILATORS

Unit ventilators, shown in Figs. 334 and 335, bring in air for ventilation and cooling as well as heating. Unit ventilators bring in air from the

outside or reuse the air already in the system. Heat or cold may be increased or decreased. The heating medium may be steam or hot water. Fans, which may be either a propeller type or a type depending on centrifugal force, shown in Figs. 334 and 335, blow air into the heater or draw the air through the heater. Schools, meeting rooms, offices, or other places that become crowded will generally be ventilated by such units. The draftsman must draw the equipment, indicate locations, and give all necessary dimensions, information, and notes.

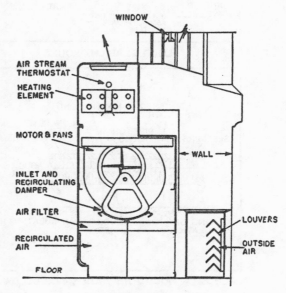

Fig. 334.
Reprinted by permission from HEATING VENTILATING
AIR CONDITIONING GUIDE, 1958, Chapter 24, p. 654.

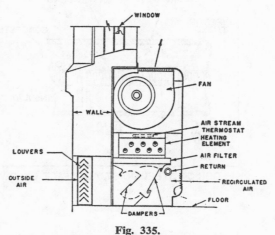
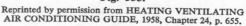

Fig. 335.
Reprinted by permission from HEATING VENTILATING
AIR CONDITIONING GUIDE, 1958, Chapter 24, p. 655.

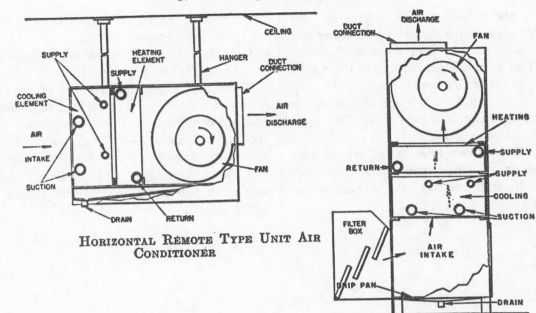

HORIZONTAL REMOTE TYPE UNIT AIR
CONDITIONER

VERTICAL REMOTE TYPE
UNIT AIR CONDITIONER

Fig. 336.

Reprinted by permission from HEATING VENTILATING
AIR CONDITIONING GUIDE, 1958, Chapter 24, p. 662.

UNIT AIR CONDITIONERS
AND AIR COOLERS

The following list, describing the various types of cooling units, has been compiled by a joint committee of engineers:[1]

A. A **cooling unit** circulates and cools air within prescribed temperature limits. Fig. 336 shows some types of air-conditioning units.

B. An **air-conditioning unit** ventilates, circulates air, cleans air, can transfer heat, and can maintain temperature and humidity within prescribed limits.

C. A **cooling air-conditioning unit** ventilates, circulates air, cleans air, can transfer heat, and can cool and maintain temperature and humidity within prescribed limits.

[1] Compiled by The American Society of Refrigerating Engineers, The American Society of Heating and Ventilating Engineers, Refrigerating Machinery Association, National Electrical Manufacturers Association, and Air Conditioning Manufacturers Association.

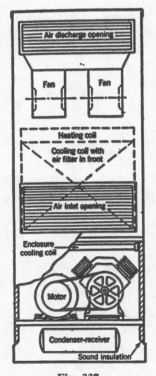

Fig. 337.

Reprinted by permission from HEATING VENTILATING
AIR CONDITIONING GUIDE, 1958, Chapter 24, p. 664.

D. A **self-contained air-conditioning unit** shown in Fig. 337, has a condensing unit besides the features mentioned above. This unit varies according to the way it gets rid of the heat created by the condenser unit, how it brings in and how it exhausts air, and how it distributes the air into the room.

E. A **free delivery tube unit** takes in air and sends it directly into the area affected, without having any outside resistance to the air.

F. A **pressure type unit** is used with elements that create air resistance.

G. A **forced circulation air cooler** transfers heat from air to refrigerants.

These are all brief descriptions of the systems and units involved in heating, ventilating, and air conditioning. For further study the *Heating Ventilating Air Conditioning Guide,* published annually by the American Society of Heating, Ventilating and Air-Conditioning Engineers, will prove an excellent source of information.

ELECTRICAL AND ELECTRONIC DRAWINGS

Although similar drafting standards and techniques are used in electrical and electronic drawings, electrical and electronic drawings differ in the material they contain.

The American Standards Association Standard C 42.70–1957, *Definitions of Electrical Terms,* published by the American Institute of Electrical Engineers, defines the noun **electronics** as "that field of science and engineering which deals with electron devices and their utilization." An **electron device** is defined as "a device in which conduction of electrons takes place through a vacuum, gas, or semiconductor."

The word **electric** is defined in The American Standards Association Standard C 42.95–1957, published by the American Institute of Electrical Engineers, as "containing, producing, arising from, actuated by, or carrying electricity, designed to carry electricity and capable of so doing." In general, electrical drawings will show electricity passing through wires.

Fig. 338 is an architect's drawing of the wiring in a house. Fig. 339 explains and defines the symbols used in the drawing.

Fig. 340 illustrates an electronic drawing. Very often electrical and electronic materials and devices will be found in the same drawing.

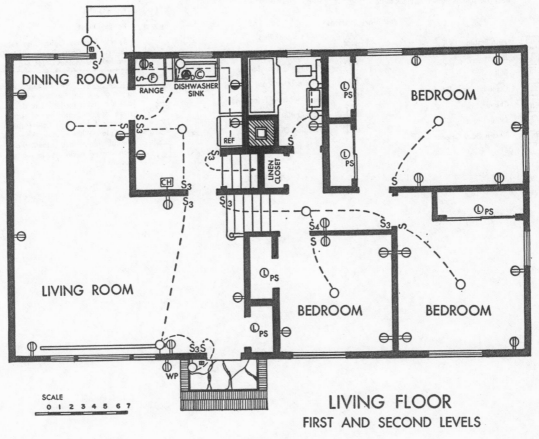

LIVING FLOOR
FIRST AND SECOND LEVELS

Fig. 338.

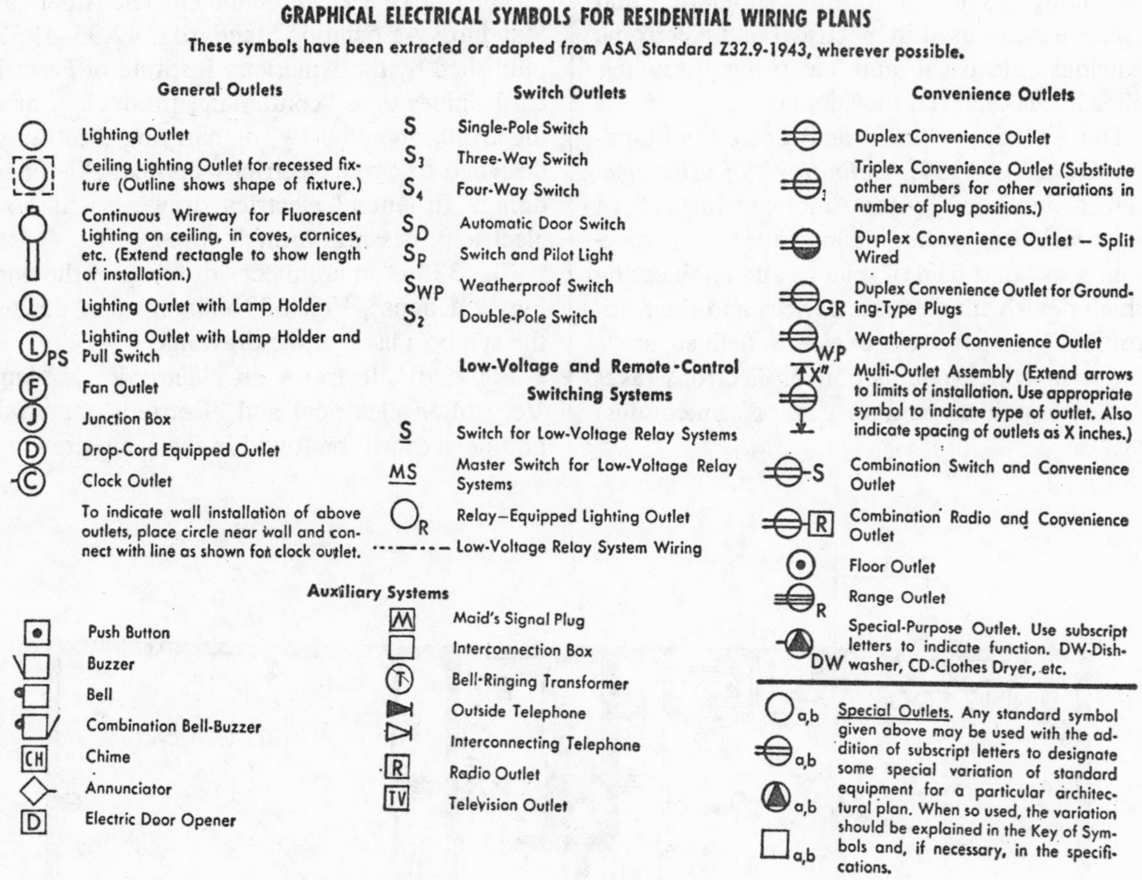

GRAPHICAL ELECTRICAL SYMBOLS FOR RESIDENTIAL WIRING PLANS

These symbols have been extracted or adapted from ASA Standard Z32.9-1943, wherever possible.

General Outlets

Lighting Outlet

Ceiling Lighting Outlet for recessed fixture (Outline shows shape of fixture.)

Continuous Wireway for Fluorescent Lighting on ceiling, in coves, cornices, etc. (Extend rectangle to show length of installation.)

Lighting Outlet with Lamp Holder

Lighting Outlet with Lamp Holder and Pull Switch

Fan Outlet

Junction Box

Drop-Cord Equipped Outlet

Clock Outlet

To indicate wall installation of above outlets, place circle near wall and connect with line as shown for clock outlet.

Auxiliary Systems

Push Button

Buzzer

Bell

Combination Bell-Buzzer

Chime

Annunciator

Electric Door Opener

Maid's Signal Plug

Interconnection Box

Bell-Ringing Transformer

Outside Telephone

Interconnecting Telephone

Radio Outlet

Television Outlet

Switch Outlets

S Single-Pole Switch

S_3 Three-Way Switch

S_4 Four-Way Switch

S_D Automatic Door Switch

S_P Switch and Pilot Light

S_{WP} Weatherproof Switch

S_2 Double-Pole Switch

Low-Voltage and Remote-Control Switching Systems

S Switch for Low-Voltage Relay Systems

MS Master Switch for Low-Voltage Relay Systems

O_R Relay—Equipped Lighting Outlet

– – – – – –   Low-Voltage Relay System Wiring

Convenience Outlets

Duplex Convenience Outlet

Triplex Convenience Outlet (Substitute other numbers for other variations in number of plug positions.)

Duplex Convenience Outlet — Split Wired

Duplex Convenience Outlet for Grounding-Type Plugs

Weatherproof Convenience Outlet

Multi-Outlet Assembly (Extend arrows to limits of installation. Use appropriate symbol to indicate type of outlet. Also indicate spacing of outlets as X inches.)

Combination Switch and Convenience Outlet

Combination Radio and Convenience Outlet

Floor Outlet

Range Outlet

Special-Purpose Outlet. Use subscript letters to indicate function. DW-Dishwasher, CD-Clothes Dryer, etc.

Special Outlets. Any standard symbol given above may be used with the addition of subscript letters to designate some special variation of standard equipment for a particular architectural plan. When so used, the variation should be explained in the Key of Symbols and, if necessary, in the specifications.

Fig. 339.

Electrical and electronic drawings, actually called **diagrams,** show how electrical energy or electrons travel from one electrical device to another. These devices contain objects, such as coiled wires in a transformer, which make some changes in the electrical energy. The changed electrical energy then travels farther to other devices until it is changed into the form of energy needed for a particular purpose. Electrical energy may be converted to heat, light, or mechanical energy. For example, the electrons going through various devices in a high-fidelity system finally end up in the speaker as a source of power which is converted to sound. In the case of an electric light bulb, the power is converted to light.

The individual interested in drawing electrical diagrams should know how the more frequently used devices affect the electrical energy or electrons passing through. He should also know the parts which each of these devices contain.

Electrical energy travels from one device to another through wires. The draftsman should learn something about the kinds of wire, how they can be connected and combined with others, what wires are used for specific purposes, and other pertinent information. A manufacturer's catalogue will provide such information.

TYPES OF DIAGRAMS

Fundamentals of Electric Wiring for the Home or Industry. The diagram that shows how electrical energy or electrons go from one device to another through wires is called a **single** or **one-line diagram.** It contains the devices, shown by standard symbols, and the wire, shown by a single line. Fig. 340 is a typical single-line electronics diagram. The complete voyage of the electrical energy or electrons is called a **circuit.** A single-line diagram may contain a system of circuits if necessary.

To show additional information, a diagram illustrating the way certain specific circuit arrangements work may be drawn. It will not in-

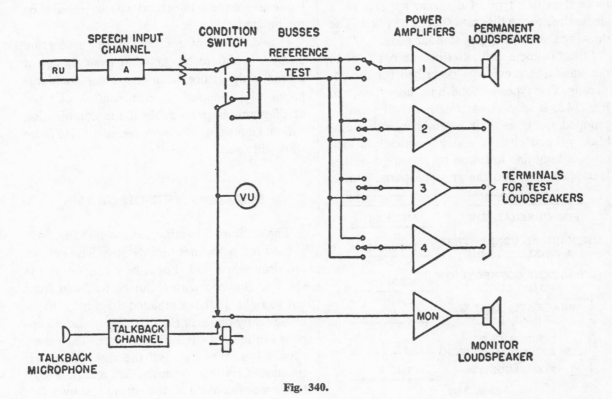

Fig. 340.

clude the physical shape, size, or location of the devices. Fig. 341 illustrates a typical **schematic** or **elementary diagram,** as this type of drawing is called.

For a still more complete picture a diagram may be drawn to show the actual connections between and to devices. All the required details of the outside and inside connections appear on the diagram. The diagram also includes the actual locations of the devices. This type of drawing is called a **connection** or **wiring diagram.**

When several large units, each containing many devices, must be connected, a diagram showing only the outside connecting lines, called an **interconnection diagram,** is drawn.

PRACTICES FOR ALL DIAGRAMS

Although the four types of diagrams just discussed serve different purposes, certain information and drawing practices apply to all of them. The title should define the type of diagram; for example, a Single-Line Diagram. If more than one type of diagram appears on a single sheet, each diagram should be titled. The drawing sheet can be any standard size.

Line thickness and lettering can conform to the standards previously described for dimensioning. For types of solid and dash lines, see Fig. 342. If several parts of a circuit must be grouped, such as unit assemblies, subassemblies, printed circuits and hermetically sealed units, a long line followed by two short dashes is used to surround the grouped parts. This is

LINE APPLICATION	LINE THICKNESS
FOR GENERAL USE	MEDIUM
MECHANICAL CONNECTING & SHIELDING LINE	MEDIUM
BRACKET CONNECTING DASH LINE	MEDIUM
BRACKETS, LEADER LINES, ETC.	THIN
BOUNDARY OF MECHANICAL GROUPING	THIN
FOR EMPHASIS	THICK

Fig. 342.

shown in the upper right-hand portion of Fig. 341.

Use the standard electrical symbols of the American Society of Mechanical Engineers (A.S.M.E.) contained in American Standard Y32.2, approved by the American Standards Association, in all your electrical drawings. The size of the symbols may vary, but they should be in proportion to the sizes of the other symbols on the diagram. The A.S.M.E. recommends a size 1½ times larger than that used in the Standard Y32.2. Any special symbols should be explained by a note.

Abbreviations should conform to the *American Standard Abbreviations for Use on Drawings,* Z32.13, published by the A.S.M.E. and approved by the American Standards Association. In the absence of specific abbreviations in the above book, abbreviations used by the electrical industry may be used. Special abbreviations should be explained.

Enough space should be provided between lines and symbols for notes. Since a great deal of blank space appears on a diagram, an even balance between lines and spaces should be maintained.

If additional information makes a diagram clearer, it should be included even though it means a departure from the usual standards previously discussed. A combination of types of diagrams is permissible if the combination will make the diagram more meaningful to those who will use it.

DRAWING THE DIAGRAMS

The medium-thick line, as shown in Fig. 342, is used for both lines and devices. The devices are drawn first and should be arranged on the sheet so that the wiring can be followed from left to right. If the completed diagram is to include many devices, the first device should appear in the upper left-hand corner. The others should follow along until the right margin is reached. Continue from the left margin below the first device until all the wiring is shown. See

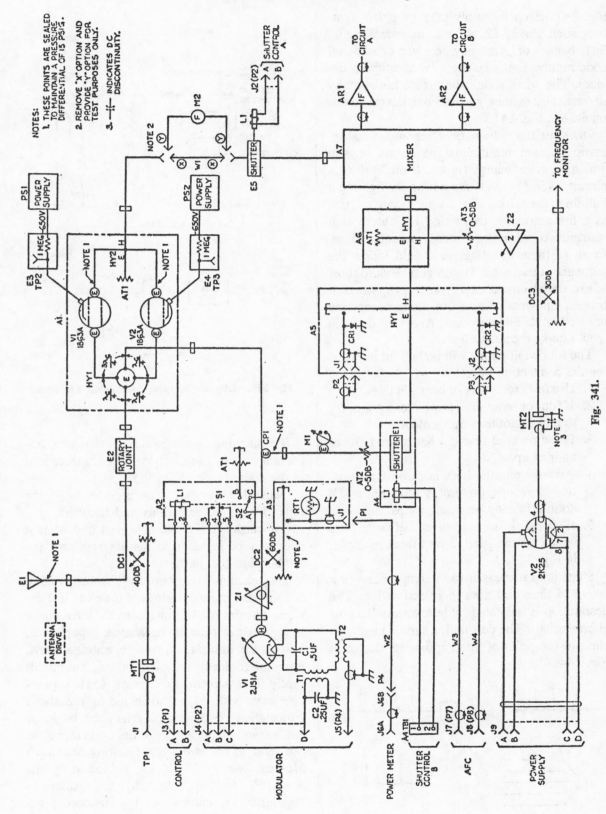

Fig. 341.

Fig. 341. Each assembly may be given numbers, such as *A1, A2, A3*, etc., as shown in Fig. 341. Names of parts may be given or national code numbers may be used for identifying devices. The wires should be drawn horizontally or vertically as they go from one device to another. See Fig. 341.

Drawing the Schematic Diagrams. A schematic diagram might show the several ways in which electrical energy travels. It might show a circuit, as in Fig. 341, the wiring for lighting a building, the wiring for a signal system, such as a fire alarm, or the wiring for automation instruments or for powerhouses and substations. In all of these the diagram should follow the sequence of the route. The diagram should show where the electrical energy comes in, where it travels, and where it goes out. Long connecting lines should not appear. Another drawing would make the story clearer.

The following points will be helpful in drawing the connecting lines:

1. Use few crossovers or bends in lines.
2. If four or more lines meet at one point, try to make another arrangement.
3. Draw parallel lines no less than 1/16 of an inch apart.
4. Arrange parallel lines according to function in groups, preferably three lines with double spacing between groups.
5. If parallel lines connect between lines which are far apart, group them as shown in Fig. 343.

Sometimes it is easier to interrupt a line or a group of lines and note their destination. The destination is indicated by letters, numbers, or abbreviations. The destination should be placed close to the point of interruption, as shown in Fig. 344.

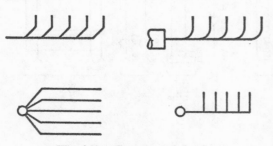

Fig. 343. Grouping of Leads.

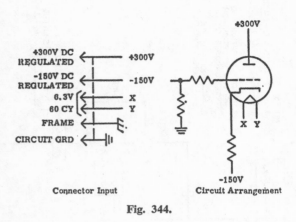

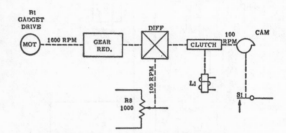

Fig. 344.

Connector Input   Circuit Arrangement

Fig. 345. Diagram Showing Mechanical Linkages.

If a machine or mechanical device is an important part of the electrical circuit, include it in the diagram. See Fig. 345.

General notes should be used to avoid the repetition of abbreviations and numbers.

For clarity, the name of a part and what it does may be included in the diagram near the symbol for the part.

The schematic diagram, as previously noted, contains the inner workings of the electrical devices. The draftsman must learn the meaning of such words as **ground, resistance, capacitance, inductance, impedance, voltage, wattage, ohms, micromicrofarad,** etc., before he can intelligently draw electrical diagrams. Certain practices exist with respect to including the above information on diagrams. When the beginner has learned the meanings of the basic terms, he can refer to the *American Drafting Standards Manual,* Section 15, Y 14, published by the A.S.M.E. and approved by the American Standards Association, for the standard practices relating to these terms.

ELECTRICAL-POWER DRAWINGS

Another type of electrical drawing might show how electrical energy is developed. **The drawing would include kilowatt rating, location of generators, how the generators are controlled, and the wiring for the controls.** If a great deal of energy is used, a special control room would be necessary. The wiring for the special control room would be put on a special drawing.

After the energy is generated it must be distributed to the points where it is needed. The power usually goes from powerhouse to substation, then to electrical instruments or other devices, sometimes overhead or underground. This distribution of electrical energy appears on interconnecting diagrams. There is a great deal of information about powerhouse and substation drawings, instrument drawings, overhead- and underground-line drawings that the draftsman must learn. He can study some of the basic practices in books, but he will acquire most of his information during his apprenticeship.

ELECTRONICS AND COMMUNICATIONS— SCHEMATIC DIAGRAMS

Schematic diagrams for electronics and communications contain many special features. Drawings of these features have been tentatively standardized by the American Standards Association. We shall study them at this point. In planning the diagram the draftsman should arrange the drawing so that it can be read from left to right, showing where the electrical energy enters, where it travels, and where it leaves, all in the proper sequence. Points where electrical energy stops for external connections should appear on the outer edges of the drawing.

The ground symbol shown in Fig. 346 appears only when the circuit ground is at a poten-

tial level equivalent to the earth potential. The ground symbol shown in Fig. 347 appears only when the connection to the housing or support of the circuit devices does not create an earth potential.

Terminals, shown by a small circle, may ordinarily be omitted, unless they make the drawing clearer. Many terminals may be required in one drawing. They may then be numbered and identified in the various ways shown in Fig. 348.

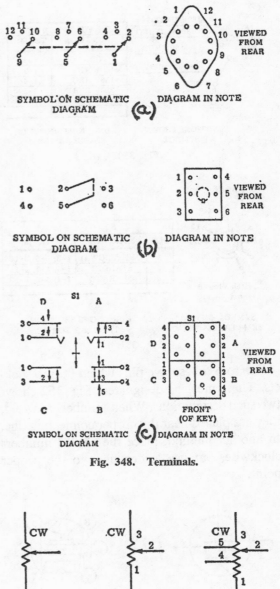

Fig. 348.   Terminals.

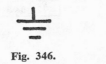

Fig. 346.

Fig. 347.

Fig. 349.

When schematic diagrams contain rotary adjustable resistors, the direction of the rotation should be shown. The letters *CW* should be put near the terminal which is first touched by the contact moving clockwise. This is shown in Fig. 349.

Switch terminals must be designated on schematic diagrams. Simple switches should have an on-off identification. Complicated switches may be shown as illustrated in Fig. 350. Rotary switches may be shown, as in Fig. 351, with a table of information added for clarity.

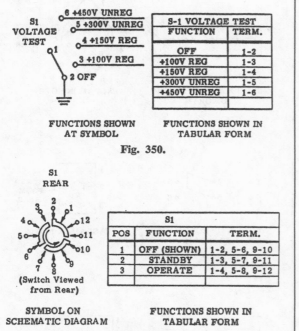

S-1 VOLTAGE TEST	
FUNCTION	TERM.
OFF	1-2
+100V REG	1-3
+150V REG	1-4
+300V UNREG	1-5
+450V UNREG	1-6

FUNCTIONS SHOWN   FUNCTIONS SHOWN IN
AT SYMBOL   TABULAR FORM

Fig. 350.

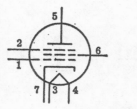

S1		
POS	FUNCTION	TERM.
1	OFF (SHOWN)	1-2, 5-6, 9-10
2	STANDBY	1-3, 5-7, 9-11
3	OPERATE	1-4, 5-8, 9-12

SYMBOL ON   FUNCTIONS SHOWN IN
SCHEMATIC DIAGRAM   TABULAR FORM

Fig. 351.

Electron tubes have pins which fit into sockets. These must be designated. Fig. 352 shows two such designations. When numbers are used, they are placed near the proper connecting line. In another method, place the circled numbers clockwise, but use leader lines to the proper point.

Fig. 352.

If a particular part has many elements, it may be shown separately on the schematic diagram for clarity.

Subdivisions of parts may be labeled with letters. If a part is designated as *Y1*, its subdivisions may be called *Y1A, Y1B*, etc. If subdivisions of a part are enclosed they may also be labeled with **suffix letters,** as shown in Fig. 353.

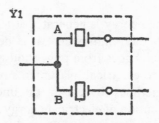

Fig. 353.   Use of Suffix Letters.

Rotary switches have several parts, each of which may be given suffix letters. Label the front of the switch as *S1A FRONT* and the back as *S1A REAR*. This is clearly shown in Fig. 354.

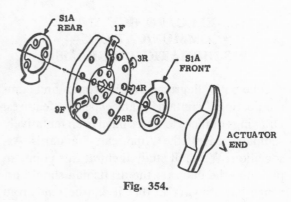

Fig. 354.

If parts of connectors, terminal boards, or rotary-switch sections work separately, show the parts separately and write the words PART OF on the diagram, as shown in Fig. 355.

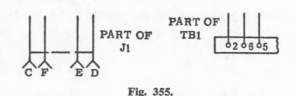

Fig. 355.

If parts of connectors or terminal boards have a great many separations, make each separation a terminal and identify each terminal separately. If individual terminals from different parts on the same drawing intermix, do not show the mechanical connecting line. Fig. 356 shows the proper identification of parts of individual terminals.

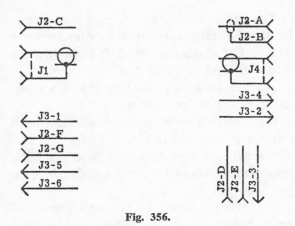

Fig. 356.

In schematic diagrams for switching circuits, parts are identified at the edge of the drawing instead of at the spot where the symbol appears. This is shown in Fig. 357.

In identifying each part of the circuit in the schematic diagram, use the American Standard List of Designating Letters, Y32 (tentative). Hyphens or spaces are not used. For example, you would use *C1,S14* and not *C-1* or *S-14*. The lowest number appears at the upper left-hand corner, moves to the right, then continues on the left-hand side of the second line and so on. If parts are eliminated, do not use the same number again. A table of information, as shown in Fig. 358, may be used.

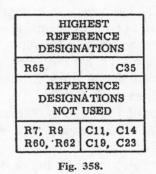

HIGHEST REFERENCE DESIGNATIONS	
R65	C35
REFERENCE DESIGNATIONS NOT USED	
R7, R9	C11, C14
R60, R62	C19, C23

Fig. 358.

Electron tubes will be identified by a designation and tube-type number. Below the tube-type the circuit function of the tube may also be indicated. Fig. 359 shows the proper reference designation for electron tubes.

Use the fewest ciphers possible in giving values of resistance, capacitance, and induct-

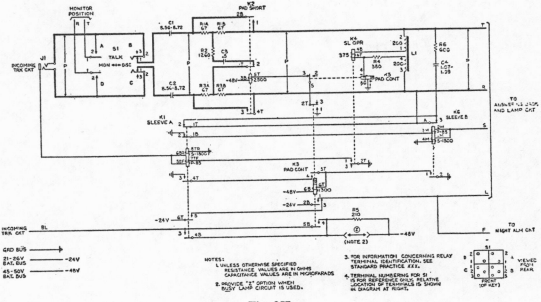

Fig. 357.

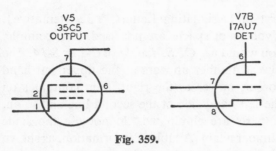

Fig. 359.

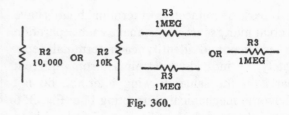

Fig. 360.

ance. In numerical values omit the coma. See Fig. 360. The basic value in expressing resistance is **ohms**. For values beginning with 1 and going to 999 the term used is **ohms. For ohms 1,000,000 and over, use megohms.** One **megohm** equals **1,000,000 ohms. 2,000,000 ohms** may be written as **2 MEG.**

For values of 1000 to 999,000, use the abbreviation *K* for **kiloohms. 56K** stands for **56,000 ohms.**

The basic value in expressing **capacitance** is **micromicrofarads.** Values up to and including 9999 are expressed as micromicrofarads. Capacitance of 10,000 micromicrofarads is expressed as **microfarads.** For example, you would write *.092 UF* (microfarads) for *92,000 UUF* (micromicrofarads).

The basic value in expressing **inductance** is **henrys (H), millihenrys (MH),** or **microhenrys (U). 2 UH stands for .002 MH and 5 MH for either .005 H or 5000 UH.**

A general drawing note is used if it is applicable to the entire drawing, as:

UNLESS OTHERWISE SPECIFIED, RESISTANCE VALUES ARE IN OHMS.

If some parts perform special functions, such information should be noted on the diagram near the proper symbol.

TEST POINT should be written on the diagram where necessary. When several such points require designation, use *TP1, TP2,* etc.

Notes should be added to the diagram whenever additional information is needed.

PIPING DRAWINGS

Many people, particularly homeowners, know that pipes carry the steam and water used in the house. Pipes also carry many other materials. Many industrial and chemical plants run mainly because the necessary materials can be carried in pipes from one vessel to another. Piping drawings show how and where these pipes should be placed. A study of standard symbols and drawing practices will make drawing or reading piping blueprints much easier.

Pipe may be made of steel, iron, glass, asbes- tos, bronze, brass, copper, lead, rubber, clay, aluminum, magnesium, and many other materials. The type of pipe material used depends upon the material that must go through the pipe. If a choice is available, the cheaper material is preferred. Steel pipes are used by the majority of chemical and industrial plants. Steel pipe is available as cast or wrought and may be purchased in a variety of sizes and lengths. All pipes that are smaller than 14″ are listed by their **nominal inside diameter.** Nominal inside

NOMINAL PIPE SIZE	OUT SIDE DIAM	NOMINAL WALL THICKNESS FOR													
		SCHED. 5	SCHED. 10	SCHED. 20	SCHED. 30	STAND- ARD	SCHED. 40	SCHED. 60	EXTRA STRONG	SCHED. 80	SCHED. 100	SCHED. 120	SCHED. 140	SCHED 160	XX STRONG
⅛	0.405	.....	0.049	.....	.....	0.068	0.068	.....	0.095	0.095	.....	.....	.....	.....	.....
¼	0.540	.....	0.065	.....	.....	0.088	0.088	.....	0.119	0.119	.....	.....	.....	.....	.....
⅜	0.675	.....	0.065	.....	.....	0.091	0.091	.....	0.126	0.126	.....	.....	.....	.....	.....
½	0.840	.....	0.083	.....	.....	0.109	0.109	.....	0.147	0.147	.....	.....	.....	0.187	0.294
¾	1.050	0.065	0.083	.....	.....	0.113	0.113	.....	0.154	0.154	.....	.....	.....	0.218	0.308
1	1.315	0.065	0.109	.....	.....	0.133	0.133	.....	0.179	0.179	.....	.....	.....	0.250	0.358
1¼	1.660	0.065	0.109	.....	.....	0.140	0.140	.....	0.191	0.191	.....	.....	.....	0.250	0.382
1½	1.900	0.065	0.109	.....	.....	0.145	0.145	.....	0.200	0.200	.....	.....	.....	0.281	0.400
2	2.375	0.065	0.109	.....	.....	0.154	0.154	.....	0.218	0.218	.....	.....	.....	0.343	0.436
2½	2.875	0.083	0.120	.....	.....	0.203	0.203	.....	0.276	0.276	.....	.....	.....	0.375	0.552
3	3.5	0.083	0.120	.....	.....	0.216	0.216	.....	0.300	0.300	.....	.....	.....	0.438	0.600
3½	4.0	0.083	0.120	.....	.....	0.226	0.226	.....	0.318	0.318	.....	.....	.....	.....	.....
4	4.5	0.083	0.120	.....	.....	0.237	0.237	.....	0.337	0.337	.....	0.438	.....	0.531	0.674
5	5.563	0.109	0.134	.....	.....	0.258	0.258	.....	0.375	0.375	.....	0.500	.....	0.625	0.750
6	6.625	0.109	0.134	.....	.....	0.280	0.280	.....	0.432	0.432	.....	0.562	.....	0.718	0.864
8	8.625	0.109	0.148	0.250	0.277	0.322	0.322	0.406	0.500	0.500	0.593	0.718	0.812	0.906	0.875
10	10.75	0.134	0.165	0.250	0.307	0.365	0.365	0.500	0.500	0.593	0.718	0.843	1.000	1.125	.....
12	12.75	0.165	0.180	0.250	0.330	0.375	0.406	0.562	0.500	0.687	0.843	1.000	1.125	1.312	.....
14 O.D.	14.0	.....	0.250	0.312	0.375	0.375	0.438	0.593	0.500	0.750	0.937	1.093	1.250	1.406	.....
16 O.D.	16.0	.....	0.250	0.312	0.375	0.375	0.500	0.656	0.500	0.843	1.031	1.218	1.438	1.593	.....
18 O.D.	18.0	.....	0.250	0.312	0.438	0.375	0.562	0.750	0.500	0.937	1.156	1.375	1.562	1.781	.....
20 O.D.	20.0	.....	0.250	0.375	0.500	0.375	0.593	0.812	0.500	1.031	1.281	1.500	1.750	1.968	.....
24 O.D.	24.0	.....	0.250	0.375	0.562	0.375	0.687	0.968	0.500	1.218	1.531	1.812	2.062	2.343	.....
30 O.D.	30.0	.....	0.312	0.500	0.625	.....	.....	.....	.....	.....	.....	.....	.....	.....	.....

All dimensions are given in inches.
The decimal thicknesses listed for the respective pipe sizes represent their nominal or average wall dimensions. The actual thicknesses may be as much as 12.5% under the nominal thickness because of mill tolerance. Thicknesses shown in light face for Schedule 60 and heavier pipe are not currently supplied by the mills, unless a certain minimum tonnage is ordered.

*Thicknesses shown in *italics* are for Schedules 5S and 10S, which are available in stainless steel only.
†Thicknesses shown in *italics* are available also in stainless steel, under the designation Schedule 40S.
§Thicknesses shown in *italics* are available also in stainless steel, under the designation Schedule 80S.

Table VI. Dimensions of Seamless and Welded Steel Pipe.

diameter is a general term and it may vary depending on the size of the pipe. The nominal inside diameter is slightly smaller than the actual inside diameter in standard pipe and is larger than the actual inside diameter in extra-strong and double-extra-strong pipe. All pipes that are larger than 12″ are listed by their outside diameter and wall thickness. This kind of pipe is known as **OD** pipe.

These thicknesses are called schedule numbers. The most commonly used thicknesses are known as **standard, extra-strong,** and **double-extra-strong pipe.** Table VI shows the nominal pipe sizes and their available wall thicknesses. The greater the pressure exerted against the inside of the pipe, the thicker the pipe will have to be.

Iron pipe is designated as **cast, malleable,** or **wrought.** When the material going through the pipe exerts low pressure against the pipe, iron pipe is generally used. Cast-iron pipe may be used for water or gas distribution systems in cities and towns. Cast-iron pipes, unlike steel pipes, have their nominal pipe sizes based upon their actual inside diameter.

Malleable and wrought cast-iron pipes are used instead of steel pipe when the material in the pipe will not corrode the cast-iron pipe. Iron pipe costs less than steel pipe. Pipe bought at a local plumbing shop is usually called **galvanized** or **black-iron pipe.**

Many of the materials used for pipe, other than steel or iron, require special study. Again, the material going through the pipe, the quantity of pipe required, and the cost must all be considered.

Tubing is also used to carry various materials. Brass and copper tubes often carry water, gas, and fuel oil used in private homes.

Copper tubing, which is available in hard and soft forms, is also used in refrigeration and air-conditioning systems. Tubing generally has much thinner walls than pipes. If your purposes require that the tubing be bent in different directions, soft tubing should be used.

PIPE FITTINGS

Various shapes are used to connect pieces of pipe or to change the direction in the piping system. These shapes are called **pipe fittings.** Means of attaching these fittings to the pipe vary. They may be welded, bolted, screwed, brazed, soldered, or packed to the pipe. The type of pipe used and the pressure in the pipe will help determine what type of fitting will be used. Special fittings can be made, but they are expensive.

An **elbow** is used to change the direction of the pipe. Elbows are designated by different types and names. The designation depends upon such things as space, connections, angles, and directions. See Figs. 361, 362, and 363.

To turn pipes 180°, **return bends** are used. Two long-radius elbows put together make a **long-radius return bend**, and two short-radius elbows put together make a **short-radius return bend.** Both the long-radius return bend and the short-radius return bend are made as one piece. See Figs. 361, 362, and 363.

One pipe sometimes comes straight out of another. A **tee,** shown in Figs. 361, 362, and 363, makes this possible. A **cross** has four openings to allow crossing of pipes. The size of a pipeline is changed by welding one size of pipe to one end of a reducer and the other size of pipe to the other end. To keep the flow of material on

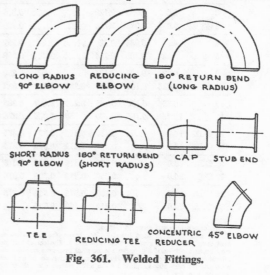

Fig. 361. Welded Fittings.

the same level, an eccentric reducer is used. Different size reducers can be obtained.

WELDED FITTINGS

Welded fittings, usually made of cast steel, are, of course, welded to pipe. Welding techniques should conform to the standards set by the American Standards Association. Welded fittings are used where strength is important. A cap is welded to the end of the pipe in order to seal the pipeline. Fig. 361 shows the types of standard welded fittings. Other welded fittings are available for various special uses. Manufacturers' catalogues will supply the necessary description and information about them.

SCREWED FITTINGS

Screwed fittings are generally used for low-pressure materials, for easy maintenance, and to save money. If necessary, screwed fittings can be obtained in steel if high-pressure materials are going to flow through the pipe. The number of threads, the length of the fitting that actually screws on to the pipe, the inside diameter, and the over-all length of the fitting have all been standardized for each pipe size. Fig. 362 shows the common types of screwed fittings.

The shapes of screwed fittings conform to the shapes of welded fittings, except for some additions, as shown in Fig. 362. A **street elbow** has threads on the outside at one end and on the inside at the other end. A **coupling** connects two pieces of pipe. There are also **plugs, bushings, unions,** and **nipples.** Nipples are very short pieces of pipe up to 12 inches long, with threads either on the inside or outside. Unions also tie two pieces of pipe together. Bushings make for easy changing from one size of pipe to another. Plugs close up ends of pipes.

A great many more screwed fittings are available in various materials. Manufacturers' catalogues illustrate them very well.

FLANGED FITTINGS

Flanged fittings are used when frequent dismantling takes place or when welding is too expensive. Flanged fittings can be made of any material used for making pipe. Flanged-fitting materials will almost always conform to the pipe material used. Cast-iron fittings will be used with cast-iron pipes. Nuts and bolts hold the flanged fittings together. All manufacturers conform to the standard for the number and size of holes, the size of the bolts, and their position on the flanged surface. These numbers and sizes will vary with the material used for the fitting. Brass fittings may have smaller and fewer holes than steel fittings. Flanged fittings made of malleable iron are shown in Fig. 363. The standard shapes of flanged fittings conform to those of welded fittings.

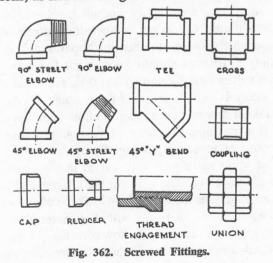

Fig. 362. **Screwed Fittings.**

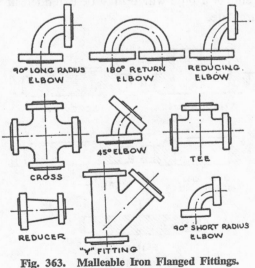

Fig. 363. **Malleable Iron Flanged Fittings.**

FLANGES

Flanges connect two pieces of pipe. They are basically thick, round disks of metal, with a large hole in the center through which the pipe goes. Smaller holes appear in a circle around the center hole. Bolts go into these holes to fasten the two flanges, and therefore the circle of small holes is called the bolt circle. Fig. 364 illustrates a bolt circle.

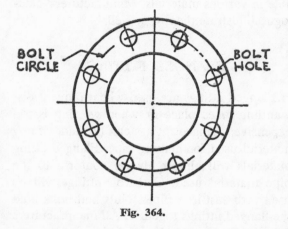

Fig. 364.

A flange may be attached to a pipe by welding, by screwing it on to the pipe, or by slipping it over the pipe end and then welding it to the pipe. The designation of the flange depends on the method of attaching it to the pipe. For example, a weld-neck flange is butted against the pipe end and welded to it. Fig. 365 shows several types of flanges. A blind flange is used at the end of a pipe which may be extended at a future date.

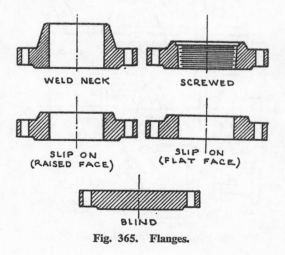

Fig. 365. Flanges.

Flanges and flange fittings have been designed to withstand different pressures from material inside the pipe. The flange that is strong enough for a particular pressure is used. Flanges have been classified for use with 150, 300, 400, 600, 900, 1500, and 2500 pounds of pressure to the inch. The draftsman must note the pressure when describing the flange. Bolts, bolt holes, and bolt-circle dimensions vary with pressure size. Flange and flange fittings with the same size and pressure have the same bolt, bolt-hole, and bolt-circle dimensions.

GASKETS

Bolting two flanges or flanged fittings will not prevent leakage. Rubber, asbestos, or synthetic rings called **gaskets** are placed on each flanged surface before bolting. A raised surface on the flange is very often provided for these rings. The piping designer usually determines the type of gasket needed. The draftsman must note the thickness of the gasket and include it when computing the over-all dimensions.

VALVES

Valves govern the flow of material through the pipes. Most of them are hand-operated, although large industrial plants use a great many automatic or electrically operated valves. In many homes, too, room temperature affects a valve which governs the heating apparatus. The piping draftsman should know what the common valves look like and what they are called. A **gate valve** either shuts off the passage of material entirely or opens it up to allow completely free passage. A **globe valve** regulates the amount of material passing through the pipe. Most water and steam valves used in the home are globe valves. The valve in the kitchen sink, which regulates the water flow, is a type of globe valve. A **check valve** prevents the material in the pipe from reversing its flow direction. A **relief valve** opens by itself when the pressure in the pipe reaches a certain level. A **control valve** regulates the flow of material automatically. It works when material in the pipe reaches a certain level, or when too much pressure builds up, and for many other purposes. A gate valve is used on either side of the control valve, and a line with a globe valve in it is run around the control valve, as shown in Fig. 366. The line around the control valve is called a **bypass.** When the control valve does not work, or should not be in operation, the bypass allows the material in the pipe to keep moving.

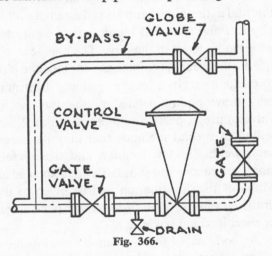
Fig. 366.

Many other valves appear in a piping system. The piping draftsman will see **angle, needle, butterfly, drain, sample, diaphragm,** and still other valve names. Experience and study will teach him when and where these valves are used.

VESSELS

Pipes carry materials from one vessel to another. A short piece of pipe, usually 4 to 6 inches long, with a flange welded to it, called a **nozzle,** is welded to the vessel. The short piece of pipe is called a **nozzle neck.** The pipe to be connected to the nozzle has a flange welded to it which matches the one on the nozzle. If the nozzle size is larger than the pipe size, which frequently happens, the draftsman must decide the easiest and cheapest way to make the transition from nozzle size to pipe size. Working with Fig. 367, if the nozzle size is 6 inches and the pipe size is 4 inches, the draftsman may first put

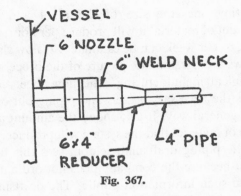
Fig. 367.

a 6-inch welding neck flange against the nozzle flange, he then places a 6- by 4-inch welding reducer against the welding neck flange, then a 4-inch pipe is placed against the reducer. Fig. 368 shows an easier and less expensive way to accomplish the same thing. Use a 6-inch blind flange with a 4-inch hole drilled into it. This is welded to the 4-inch pipe and bolted to the 6-inch nozzle.

The draftsman will encounter many types of vessels, often requiring special pipe fittings. Experience will help him solve many of these problems.

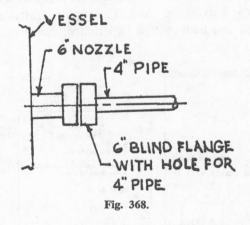
Fig. 368.

USE OF PIPING DRAWINGS

The piping draftsman might be interested in knowing how his work fits into the whole process of designing and building a plant. Let us use a chemical plant, which requires many piping drawings, as an example. Research in the laboratory develops the basic process for manufacturing a chemical product. After a company decides to build a new plant or to enlarge an existing one, the size of the plant and the amount of material it will produce per hour, per day, or per week can be estimated. A **flow sheet** shows a diagrammatic picture of the process of chemical manufacturing. The flow sheet contains the necessary vessels and piping but only in a general way. It also shows the amount and type of material that the system will produce.

The piping draftsman seldom sees this flow sheet because the company does not ordinarily make such information public. The draftsman will see **flow diagrams,** such as the one shown in Fig. 369. Although flow diagrams contain more precise information as to vessel and pipe sizes, they are diagrammatic. The structural, architectural, concrete, instrument, civil, and heating and ventilating design departments will be working on their particular phases of the chemical plant and will submit their drawings to the piping department as they finish them. The piping draftsman will work on one small part of the building. Usually many buildings make up the entire plant and many draftsmen must be employed. The piping engineers acquainted with the entire chemical process locate the vessels in the plant. The piping draftsman, if he has had enough experience, can arrange

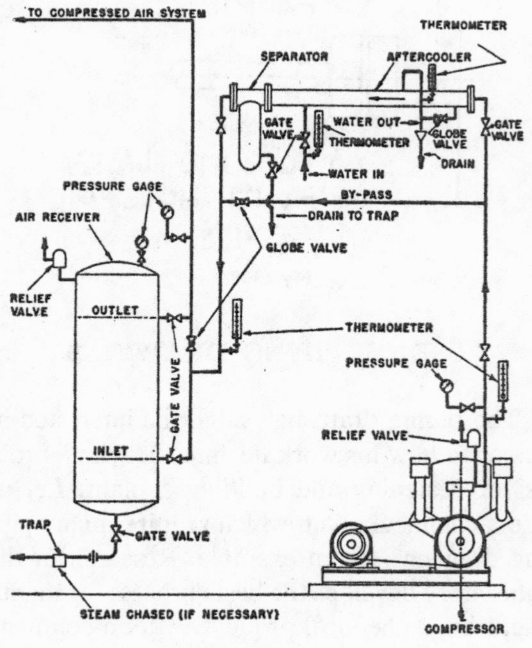

Fig. 369.

the piping from one vessel to another, and from one pipe to another, following the diagrammatic arrangement on the flow diagram.

The piping arrangement must be clear of the steel in the building, of walls, stairs, platforms, or any other interfering objects. The piping draftsman uses the drawings of the other departments to learn what interferences may exist. The pipe may be arranged so that it can be supported; so that the valves may be reached easily; so that the least amount of fittings are required; so that workmen will not be injured while working under the pipes, and to satisfy many other physical and chemical requirements connected with the material in the pipe. The engineer will tell the draftsman, for instance, to slope the pipe to create a gravity effect. The piping draftsman may draw piping details, noting the special valves, fittings, parts, and pipes required. Notes telling of special methods that may be necessary, welding notes, location and size dimensions, references and standards are also needed. Much of this information will be given to the draftsman in the form of mimeographed pages or specification booklets.

A good deal of the draftsman's work will involve changes that the piping engineer may make with respect to design or arrangement. Companies first issue preliminary drawings. After study by the proper personnel, the drawings are returned for further work and then, containing the signature of the supervising engineer, are issued as final. The construction crew in the field then receives them and starts to build. Construction usually starts with pouring the concrete, then putting up the steel, and then placing the vessels. The piping may still be subject to change during the initial phases of construction. If the draftsman changes the piping after the final issue, he must note these revisions on the drawing. Each company has its own method of noting such revisions.

One of the most important problems which faces the draftsman, besides the actual drawing, is the gathering of necessary information. He must wade through all the blueprints affecting his own drawing in order to know all about the part of the plant on which he is working. The draftsman must study all manufacturers' catalogues which list equipment used on the pipes that he must draw. The draftsman must refer to company standards for assembling piping parts. He must also ask questions of superiors and be able to draw properly.

MAKING A PIPING DRAWING

The piping draftsman will usually work on only a small portion of the entire building. In making his drawings the draftsman will put plan views on one sheet of paper, and elevations and sections will be placed on other sheets of paper. The center lines of the steel beams and columns should be put on first, in light lines, with a 2H pencil. The columns are the vertical steel members. They are drawn to scale in solid black. The column sizes are obtained from a list called the **column schedule** and are provided by the steel draftsman. The vessels and other equipment should then be placed where they belong. Fig. 370 shows a typical vessel drawing. Center lines of equipment pieces are drawn before the equipment itself is drawn. The equipment-design department usually provides all the necessary dimensions for the vessels, nozzles, pumps, or any other equipment used. As shown in Fig. 370, nozzles must be placed in their proper location. Light lines that are somewhat heavier than center lines should be used for equipment.

If the draftsman designs the piping for his area, he usually makes a rough sketch first. He should take into account all the problems previously discussed. If the draftsman is required to draw a piping layout given to him, he should draw all center lines of pipes first. In order to save time, many companies use single lines for pipe sizes up to 10 or 12 inches and double lines for larger pipes. After the draftsman draws the center lines he locates and draws the fittings and valves. The standard symbols for fittings and valves, shown in Fig. 371, have been approved by the American Standards Association and published by the American Society of Mechanical Engineers. In some cases certain symbols may vary according to usage by individual companies. The pipe should now be drawn. A line heavier and darker than those used for steel equipment and center lines should be used. Lines for elbows and flanges should be as dark and heavy as pipelines. The crosslines of valve symbols are, however, lighter.

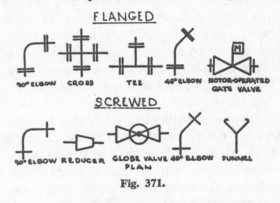

Fig. 371.

DIMENSIONING A PIPING DRAWING

Locating Dimensions. Study Fig. 372. First give the dimensions from steel column to steel column. Then locate all pieces of equipment from center lines of columns to center lines of equipment. Nozzles are located on vessels by giving the dimensions from the center line of the vessel to the face of the nozzle and by including the angle the nozzle center line makes with the vessel center line. See Fig. 372. Any other special situations that require locating dimensions should, of course, be included.

Pipe-locating dimensions run from center lines of columns to center lines of pipe or from one pipe center line to another pipe center line, as shown in Fig. 373. Pipes are never located from their outside surface. When pipes change

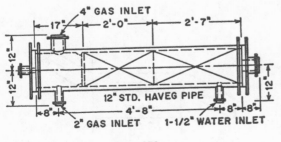

Fig. 370.

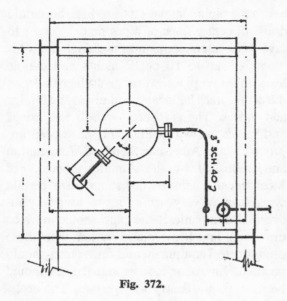

Fig. 372.

directions, the location of the turning point is given from a column center line or pipe center line, as shown in Fig. 373. All locating dimensions appear on the plan view only. Elevation dimensions appear on elevation drawings only. This rule may be broken in special situations.

Valves should be located when their location point is important. Control-valve arrangements or other special combinations of valves and fittings should have locating dimensions.

Architectural features, such as stairs, doors, platforms, and walls, which might affect piping arrangements should be included and located in the drawing. This also holds true for heating and ventilating systems, concrete, plumbing, and electrical conduits. These items are drawn in light dotted lines to avoid confusion with the lines drawn for pipe.

SIZE DIMENSIONS

Straight pieces of pipe are dimensioned from one center line to the center line of the fitting which turns the pipe in another direction. Pipe is purchased in standard lengths. When the design calls for long runs of pipe, for example 60 feet, if the pipe is purchased in 20-foot lengths, flanges are inserted at 20-foot intervals. The pipe could be welded into one piece, but that would make it difficult to dismantle. The length of the pipe is marked from the face of one flange to the face of the flange on the other end of the pipe, as shown in Fig. 374. The true lengths of the pipes which slope or appear foreshortened are given. The draftsman should add the words *true length,* or the abbreviation T.L., in parentheses after the dimension or under it, as shown in Fig. 375.

Standard fittings are not generally dimensioned, but special fittings do require dimensions. When a 45° elbow is used, the angle must be given, as shown in Fig. 376. Face to face dimensions of valves may sometimes be required.

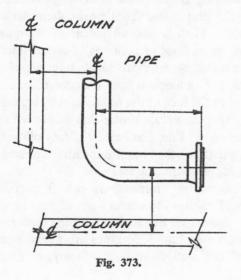

Fig. 373.

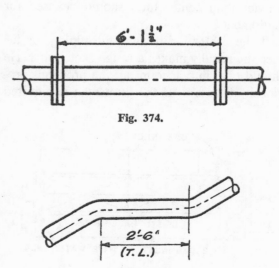

Fig. 374.

Fig. 375.

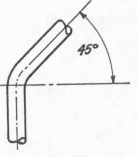

Fig. 376.

NOTES

Each company develops its own method for denoting the size and type of pipes, fittings, and valves and for other pertinent information. The draftsman, however, should acquaint himself with some of the more common practices. Pipe sizes have been discussed previously. The draftsman denotes the pipe size as illustrated in Fig. 372. The American Society of Mechanical Engineers has published and the American Standards Association has approved the standard nomenclature for the various kinds of pipe. If the kind of pipe must be noted in the drawing, the draftsman should use the proper American Society of Mechanical Engineers' designation. The note, for example, all pipe to be SA-53, Grade A, should be placed with any other required notes. Since a plant needs a great many pipes, they must all be marked so that they can be distinguished from each other. Each company has its own pipe-numbering system. Separate drawings of the same area of a plant are made for **process piping** and **service piping**. Process pipes will carry materials used in making the product. Service pipes will carry water, steam, air, nitrogen, and similar materials which help in producing but are not part of the final material. Similar drawing procedures are used for both types of pipes.

The type and size of fittings are noted as illustrated in Fig. 372. If all the fittings are made of the same material, a general note to that effect should be used instead of naming each fitting. The note might say, for example, all fittings to be 125※, cast iron, flanged. If the piping design calls for a fitting or valve made by a particular manufacturer, the name of the manufacturer, his catalogue number, the particular page, and the fitting or valve number must be given.

Many additional types of notes relating to piping designs and dimensions will probably be needed. The draftsman will be told or be given standards to follow to make his drawing complete.

Draftsmen may sometimes receive piping information in the form of isometrics, from which they draw plans and elevations. The piping draftsman may, on the other hand, draw a piping layout in isometric to make the picture clearer. Fig. 377 illustrates piping in isometric.

PLUMBING

Pipes carry clean water to buildings from water mains in the street and carry dirty water and waste away from the buildings to sewers. The engineer or architect determines how and where these water pipes will be located in the building. Clean water is needed for lavatories, showers, sinks, urinals, etc. Plumbing drawings show how and where the water pipes come into the building; how and where they run in the building—to the proper fixture; and how and where they carry the waste material out of the building. Plumbing drawings are very similar to other piping drawings, except that the type of pipe and pipe fitting used are specially made for plumbing. In well-built residential buildings copper tubing, with copper and brass fittings, is used. Copper tubing may be obtained in a hard-temper form, known as type *K*, or in a soft temper, known as type *L*. Type *M* is also a hard-temper tube, but comes in larger diameter sizes.

Cast-iron, bell, and spigot pipes and fittings are very widely used. Fig. 378 illustrates this type. The spigot fits into the bell end. The bell is filled with cement, oakum, lead, or rubber to prevent leakage. Bell and spigot pipes can be obtained in 12- and 16-foot lengths.

Plumbing pipe is also available in aluminum and stainless steel.

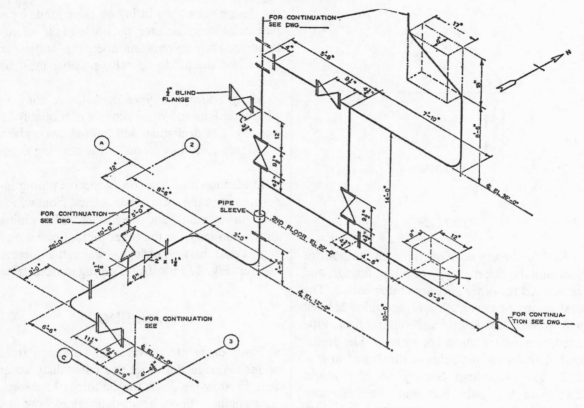

Fig. 377.

Lavatories, sinks, showers, and all other plumbing fixtures must be located and dimensioned by the draftsman. Fig. 379 illustrates symbols for some of these fixtures. Fig. 380 shows a diagrammatic arrangement of a restaurant kitchen.

The draftsman should acquaint himself with some of the plumbing parts accompanying a sink, such as the traps, faucet, faucet-hole cover, and rim guards. Each fixture has its own arrangement.

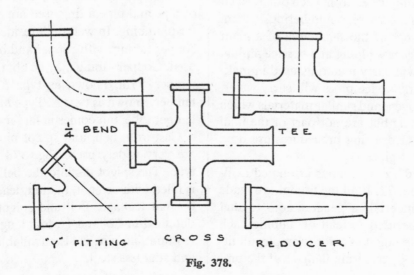

Fig. 378.

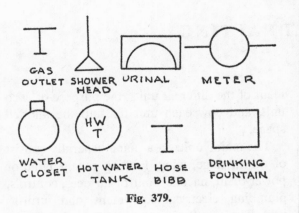

Fig. 379.

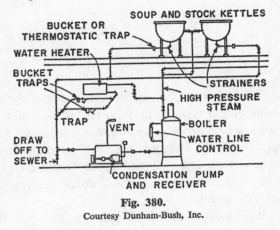

Fig. 380.
Courtesy Dunham-Bush, Inc.

Pipes always go through floors, walls, and ceilings. If these are quite thick, sleeves must be inserted. The pipe then goes through the sleeve. The sleeves, made of cast iron, cement, and ceramic, will protect the ceilings, walls, or floors from hot or cold water or steam. Pipes have plates around them. The plates rest on the floor or are attached to the wall or ceiling. They serve as further protection and as decorations when used in a home.

CHAPTER SIXTEEN

AIRCRAFT DRAFTING

Aircraft drafting requires the same skills which other types of drafting require as well as some which are unique to aircraft drafting. As you read this chapter, you will be reviewing much of what you learned about dimensioning, sections, assembly drawings, detail drawings, and diagrams.

The inexperienced aircraft draftsman will probably begin his career as a tracer or he may be asked to make simple detail drawings. In this way he will become familiar with the special drawing practices, the parts of the plane, and the special vocabulary used in the aircraft industry.

The aeronautical designer creates the idea for the entire plane. After the draftsman receives a sketch from the designer he draws it to scale. The sketch should show the arrangement of the different units, the purpose of each unit, and how each unit fits into the allotted space.

The major units in a plane generally consist of the fuselage, wing, landing gear, engine or power plant, nacelle, control surfaces, controls, plumbing, electrical equipment, and furnishings. The nacelle is an enclosed shelter on an aircraft for passengers or for a power plant. Fig. 381 shows the main parts of the plane except for the control surfaces, controls, plumbing, electrical equipment, and furnishings.

The detailer may see the arrangement or **layout** drawing of an entire plane as well as the drawings of individual units. (The drawings being discussed at this point are not parts of the plane shown in Fig. 381.) Fig. 382 shows a unit drawing of a fuselage. Fig. 383 shows an aileron

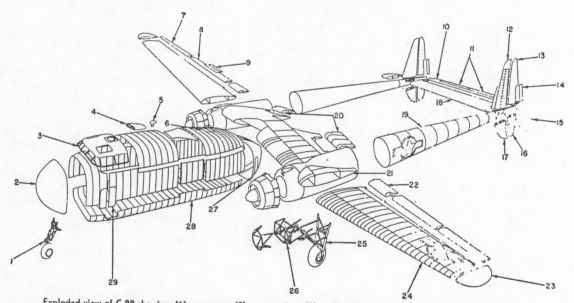

Exploded view of C-82 showing: (1) nose gear; (2) nose section; (3) cockpit enclosure; (4) scoop; (5) antenna; (6) life-raft door; (7) outboard aileron; (8) inboard aileron; (9) tab; (10) elevator; (11) tabs (spring and trim); (12) upper fin; (13) upper rudder; (14) tab; (15) stabilizer tip; (16) lower rudder; (17) lower fin; (18) stabilizer; (19) boom; (20) inboard flaps; (21) wing center section; (22) outboard flap; (23) wing tip; (24) outer panel; (25) main landing gear; (26) nacelle structure; (27) rear cargo door; (28) fuselage main body section; (29) front cargo door located below cockpit floor.

Fig. 381.

By permission from AIRCRAFT DESIGNERS' DATA BOOK by L. E. Neville. Copyright, 1950. McGraw-Hill Book Co., Inc.

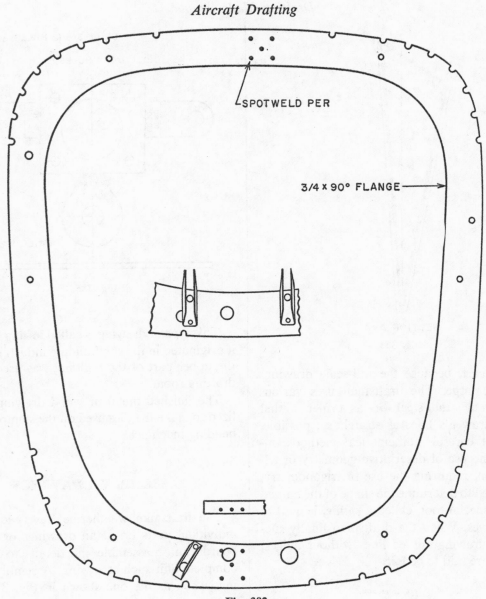

SPOTWELD PER

3/4 x 90° FLANGE

Fig. 382.

assembly. Fig. 384 shows a detail drawing of a fuselage, BLKD ⅜5. (The drawings in Figs. 382 through 386 are all from different planes.)

The detailer will acquire some perspective of where his own drawings fit into the total pattern of the plane when he sees a drawing of the entire plane and individual drawings of the component units which comprise the plane.

It is essential that the detailer have a thorough knowledge of the drafting practices used in the machine shop. As illustrated in Fig. 385, all of the drafting practices previously discussed will apply.

LOFTING

A plane has an unusual shape with many complicated curves. The unusual shape is created by using many lightweight aluminum pieces. Many types of stamping and bending machines are required to make these pieces. The drawings used in the construction of these odd-shaped pieces are made on metal, wood, or plastic by a specially trained draftsman, not a detailer, because they must be extremely accurate and drawn to full scale. The draftsman works on the floor, sometimes lying on a

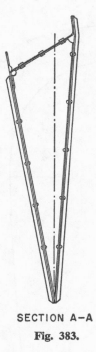

SECTION A–A

Fig. 383.

wheeled cart, because the full-scale drawings are quite large. The draftsman uses various aids, such as a **taut steel wire** as a ruler, so that he can draw his lines as straight as possible. The most careful mathematical methods, including the use of descriptive geometry or **triangulation,** requiring the use of trigonometry, are utilized to determine the shape of the curved lines. A special tool, called a **spline,** is used to draw curves. When the draftsman finally succeeds in drawing curved lines without bulges, the lines are said to be **fair.**

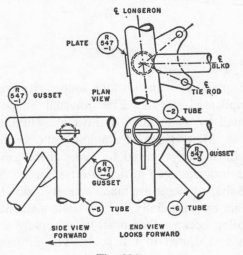

Fig. 384.

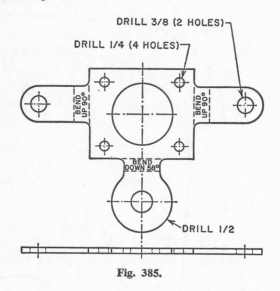

Fig. 385.

This type of drawing is called **lofting** because it originated in the shipbuilding industry, where the upper part of the building was used as the drafting room.

The finished metal or wood drawing is utilized to make the dies used on the stamping and bending machines.

ASSEMBLY DRAWINGS

The draftsman, whether he draws each of the individual parts on detail drawings or on loft boards, must assemble the details to form a complete unit such as a wing. Assembly drawings, such as the one shown in Fig. 382, are made to show how these pieces are put together. If the unit seems too complicated to be shown in one assembly drawing, smaller drawings, called subassemblies, are made to clarify the situation. To show how all the units are combined to compose the finished plane, a **major assembly** or **installation drawing** is made.

DASH-NUMBER SYSTEM

Each assembly will require a great many small parts. In order to prevent confusion and make it easier for the assemblers, a **dash-number** system is used for each part of the assembly as well as for each unit in the plane. The dash number assigned to each part appears on the original detail drawing for that part of the plane. This is shown in Figs. 384 and 386.

The dash number is simply a number preceded by a dash. Both the number and the dash are placed inside a circle. A leader, shown in Fig. 384, points to the proper part. In some instances the dash number may also include the name of the part. If the same part appears more than once on one drawing, the dash number is not repeated.

A part may appear on an assembly to which it does not belong. It is there merely to show its relationship to the other parts of the plane. In such an instance, the dash number of this reference part is given as it appears **on the detail drawing of which it is a part.** The letters REF are placed near the reference part to show that it is there for reference only. This is shown in Fig. 387.

Once a dash number is used, it is never changed. If a dash number is canceled, that number is not assigned again.

Each detail drawing has a drawing number. The dash number sometimes appears after the drawing number, as, for example, 63C467–6 (the final 6 is the dash number). This type of

Fig. 387.

designation is often found with roughly finished or unmachined castings which cannot be accurately dimensioned.

The exterior of a plane is generally symmetrical in shape. If one drawing is used to represent both sides of the plane, the letters *L* or *R,* as shown in Fig. 384, should appear near a dash number to show on which side the part actually belongs.

A table, which serves as a key to what each dash number represents, is put on the same drawing or on a bill of materials. See Fig. 384.

DIMENSIONING

The dimensioning practices used on detail drawings follow the standard practices used in dimensioning. However, the dimensioning practices used in assembly drawings require special study.

Subassembly Drawings. Subassembly drawings appear in ½-inch scale or full scale and may contain locating dimensions. A **reference center line,** which refers to an adjacent subassembly, appears on the drawing. One edge of the assembly is located from the reference center line. Each succeeding part of the assembly is located from that edge. Study Fig. 388. The mechanic putting the parts together will do so in the order shown on the drawing. The proper procedure is for the mechanic to begin with the reference center line and assemble the pieces, working away from the reference center line.

The draftsman must locate the rivet holes carefully. He must show the side on which the rivet head appears and must note the type and number of rivets required for each hole, as shown in Fig. 388. Fig. 388 also shows that the draftsman must locate the skin or outside of the plane and note how it is to be riveted to the frame.

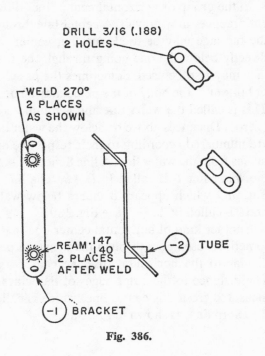

Fig. 386.

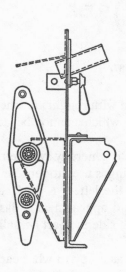

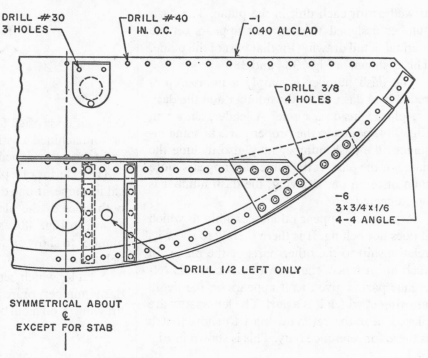

DRILL #30
3 HOLES

DRILL #40
1 IN. O.C.

−1
.040 ALCLAD

DRILL 3/8
4 HOLES

−6
3 x 3/4 x 1/6
4-4 ANGLE

DRILL 1/2 LEFT ONLY

SYMMETRICAL ABOUT
℄
EXCEPT FOR STAB

Fig. 388.

Installation Drawings. Because the installation assembly is usually quite large, a ¼-inch scale is used. Again, the draftsman, for the most part, uses locating dimensions. All the necessary reference parts appear in **phantom** or **dotted lines.** Installation assemblies must be shown in relation to the entire plane. Since the plane is large, it is divided into vertical and horizontal reference or center lines. Two groups of vertical center lines are used. One group covers the plane from left to right, across the wings, and the other group covers the plane from the front or nose to the back, as shown in Fig. 389. In the first group, the vertical center line or the fuselage is the main center line, sometimes called zero. Any vertical center lines to the left or right of this main vertical or fuselage center line are called **buttock** lines and are measured by their distance from the main vertical center line. See Fig. 389.

The second group of vertical center lines starts at the nose of the plane. The first-line station is called *0* (zero). All other center lines behind the first one are numbered according to how far back they are, usually in inches, from Station *0*. If the second vertical line is 23 inches behind Station *0*, it is called Station 23, as shown in Fig. 389.

In the group of horizontal center lines different practices may prevail. A center line through the fuselage may be used. Another center line directly below the one going through the fuselage may also be used. Sometimes the base line or bottom of the body of the plane may be used. This is called the **water line** and is numbered *0* (zero). Distances above or below the water line are numbered according to their respective distances from the water line. A line 8 inches above the water line *0* is called *W.L.+8*. The line in Fig. 389 which appears 8 inches below water line *0* is called *W.L. −8*. See Fig. 389.

Another kind of horizontal center line that is sometimes used goes through the wings perpendicular to the horizontal fuselage center line. The distance to the outer edges of the wings is measured from this center line, which is called the **chord line,** as shown in Fig. 389.

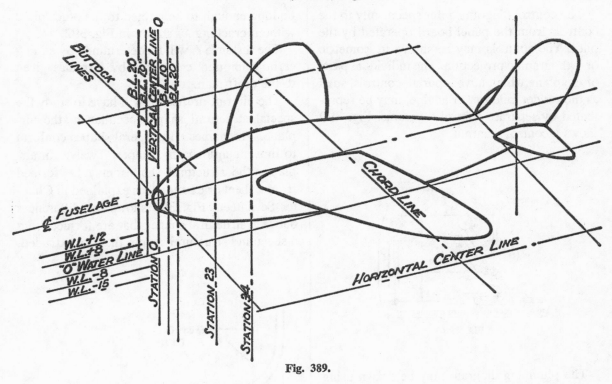

Fig. 389.

FINAL ASSEMBLY DRAWING

The final assembly drawing shows the over-all relationship of the different units which comprise the plane. Individual assemblies and installations appear with the appropriate names and numbers. This is shown in Fig. 381. The three views, the top view, the front view, and the side view, are drawn with only the over-all dimensions given. The final assembly drawing is designed to show the reader the total picture of the plane and indicates exactly where each unit belongs. It also serves as a convenient index to the entire plane, since each unit is shown with its proper drawing number. By finding each unit drawing, any interested individual can find the drawings for each subassembly and detail.

DIAGRAMS

The electrical, the control, and the plumbing systems are shown in diagrams. The plumbing system includes the hydraulic, the fuel, and the de-icing lines. In all probability the draftsman will limit himself to working in one or two of the above areas.

Electrical diagrams include the symbols and drawings studied in Chapter Fourteen. Special electrical symbols devised by individual companies may be used to augment the standard symbols. The draftsman must, of course, be familiar with the different electrical circuits and devices used on the particular type of plane on which he works. Since electrical energy runs through almost every part of the plane, it would be impossible to show the entire electrical system in one drawing, and the draftsman must make a drawing for each part of the system.

All wires must be covered, and the draftsman must remember to show shieldings in all his diagrams.

The draftsman must know how his particular company provides for necessary references on drawings. References are important to prevent the crossing of wires. Each diagram must be correlated to show where each wire enters as well as its ultimate destination.

The control diagrams refer specifically to the controls from the panel board operated by the pilot. The controls may be shown in isometric or orthographic projection. The individual parts of the plane which have separate controls, such as the rudder, aileron, or engine, may be represented by separate control diagrams. Fig. 390 shows a control diagram.

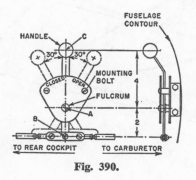

Fig. 390.

The plumbing diagram may be drawn using the methods outlined in Chapter Fifteen. The plumbing diagram may also be shown in isometric projection. The various units of the plane allied with the plumbing system, such as the hydraulic system or the fuel system, are drawn on separate diagrams.

The aircraft draftsman should make particular note of the drawing of **bends.** Many of the airplane parts must be bent. A piece of metal is bent according to a certain radius called the **bend radius.** The bend **tangent line** shows the points where the bend begins and ends. See Fig. 391. If a bent piece of metal is joined to another, the extended area of the bent piece must be long enough to be riveted to the other piece without cracking, as shown in Fig. 392.

The distance L will be determined by adding certain amounts calculated by formulas given to the draftsman.

The aircraft draftsman will have to study the special structural shapes peculiar to the airplane. The names and general shapes conform to those mentioned in Chapter Twelve. In airplanes the structural shapes may be formed from a sheet or extruded, as explained in Chapter Seventeen. Fig. 393 shows the differences between structural shapes that are formed from a sheet and structural shapes that are extruded.

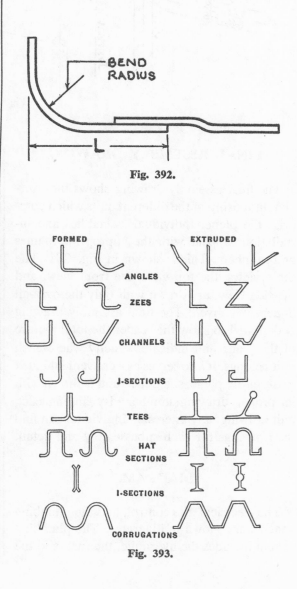

Fig. 392.

Fig. 393.

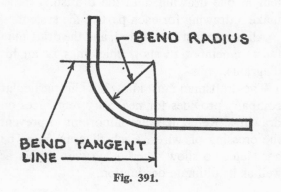

Fig. 391.

CHAPTER SEVENTEEN

PLASTICS

The number of companies manufacturing objects made of plastics has grown very large. The draftsman who wishes to devote his time to making drawings to be used in the construction of plastic objects must learn the words and symbols most commonly used in this field. The plastics industry is not very old, but in 1958 the American Society of Mechanical Engineers, with the approval of the American Standards Association, released Y 14.11, *The Plastics Drafting Standards Manual.*

As in other specialized fields, the draftsman should know something about the objects he draws. He must also know something about plastic materials, since the type of plastic used determines the way the object is made.

TYPES OF PLASTICS

Various chemicals are combined to form a new material which is called a plastic. The word plastic refers to the ability to be molded or shaped. This new material can be put into any kind of shape. To make the desired shape a negative or mold is needed. The plastic is poured or forced into the mold, and when it hardens it is removed from the mold.

If certain chemicals are used to make the plastic material, and then heat is applied to the material, the plastic will become hard and it is impossible to melt or dissolve it. This type of plastic material is called **thermosetting plastic.** Heat is applied to it by use of hot molds, which are placed in operation when pouring or pressing the plastic into the mold.

If the finished object must resist heat when used, minerals can be added to the material before pouring it into the mold. The following fillers may also be added: For electrical insulation, mica; for resistance to chemicals, inert materials; for withstanding pounding, fabrics; for general purposes, wood flour.

When thermosetting plastics are used to form sheets, **cotton fabric, linen, cellulose, paper, glass-fiber fabric, asbestos fabric, glass-fiber mat, asbestos mat,** or **wood veneer** are added. A sheet of plastic is formed, the desired material is then added, and another sheet of plastic is poured over the desired material, sandwiching it in permanently. This is called **laminate.** The material used between the thermosetting sheets depends upon the purpose of the finished object.

If other groups of chemicals are used to make the plastic material, and if heat is applied to the material, it will merely soften. This type of material is called **thermoplastic.** An object is made of a thermoplastic by heating it to a semi-fluid state and forcing or injecting it into a cool mold. The object is chilled before it is removed from the mold. Fillers are not generally used with thermoplastics.

DESIGNATING THE TYPE OF PLASTIC

Since so many varieties of plastics are made, the draftsman must note on his drawing whether the plastic is thermosetting or thermoplastic. Many companies have their own trade names or numbers. These should be on the drawing. Some plastics have been given numbers by the American Society for Testing Materials and the National Electrical Manufacturers Association. These numbers should be on the drawings when such plastics are used. If the plastic material to be used must have a certain color, go through certain electrical or mechanical tests, or be porous to a certain degree, such information should be included in the drawing.

FORMING PROCESSES

The drawing will consist mainly of the shape of the object and the shape of the mold in which the object will be made. The shape of the mold will depend, in most cases, on whether thermosetting or thermoplastic material is used, and on how the material is put into the mold. The two basic ways of putting material in a mold are known as the **pressure-shaping** method and the **casting** method. In the casting process the plastic material, which may be either thermosetting or thermoplastic, is simply poured into the mold. The molds, which are hot when thermoplastic materials are poured into them, are cooled. Cooling hardens the material, and the object is removed from the mold after hardening. Thermosetting material, on the other hand, requires heating in order to cure or harden before it is removed from the mold.

PRESSURE SHAPING

The majority of plastic objects are formed by pressing them into the desired shape. This is done by filling one half of a heated mold with thermosetting material, and pressing down the other half of the mold, forcing the material to take the shape formed by the two halves of the mold. Fig. 394 illustrates the **compression-molding principle.** The material must cure or harden to form a compression mold. This is accomplished by using the amount of heat and pressure required by each type of thermosetting material.

Thermoplastic material may also be used in a compression mold, but it takes much longer to set. Therefore, the compression-molding technique is used infrequently with thermoplastics.

INJECTION MOLDING

Instead of pouring material into an open mold, thermoplastic material can be squirted into a closed mold. This type of mold is called an **injection** mold. Fig. 395 illustrates the principle of injection molding. The process of squirting is called a shot. Many injection machines work automatically, thus allowing for great speed in production. A number of copies of the same object can be made at the same time in a cluster and can be cut from one another by machine.

TRANSFER MOLDING

When objects have intricate, fragile parts, or long, slender ones, a combination of the compression and injection methods, known as **transfer molding,** illustrated in Fig. 396, is used. Thermosetting materials are used mostly. The material is heated to its softening point in a transfer chamber and then forced through a sprue into a closed mold. Heat and pressure cure the material.

Sometimes the material is heated before it is put in the transfer chamber and then forced into the mold with a plunger. This method, which allows for greater speed, is known as **plunger molding.**

EXTRUDING

When objects have one shape, such as a sheet, rod, or tube, and are made of thermoplastic material, the **extrusion process,** illustrated in Fig. 397, is used.

The material is softened with heat and forced through an area which has the shape of the finished product. A screw forces the material through a shape called a **die.** When it comes out of the die, it is in a semisolid state and must be chilled. A conveyor belt receives the material and water is sprayed on the material to cool it. See Fig. 397.

HOT FORMING

When the desired object is large and can be made of thermoplastic material, a large sheet can be placed over the desired shape and the sheet may then be forced down by vacuum or air pressure. After the sheet cools, the formed object can be removed.

LAMINATING

Cellulose paper, cotton fabric, glass fabric, and wood veneer are types of material that are mixed in with or sandwiched between thermosetting plastic by heat and pressure.

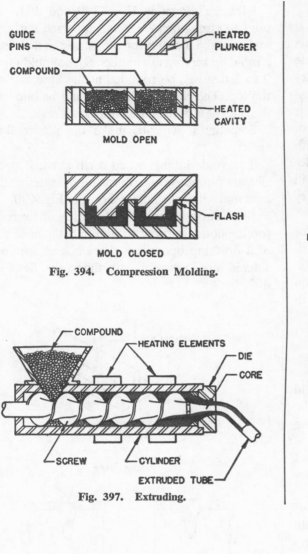

Fig. 394. Compression Molding.

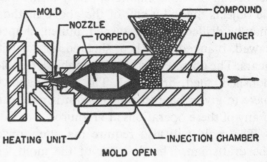

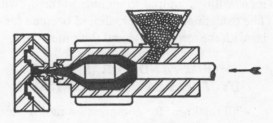

Fig. 395. Injection Molding.

Fig. 397. Extruding.

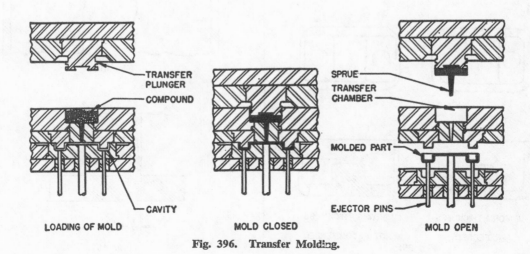

Fig. 396. Transfer Molding.

DRAWING TECHNIQUES

The draftsman must draw the object in all the necessary views and sections. He must provide all the dimensions, notes, and specifications described in previous chapters. Many of the plastic objects made by any of the above methods may be drilled, tapped, turned, milled, shaped, sawed, blanked, punched, sheared, or cut with gears. The objects may also be tumbled, tumble-polished, filed, or buffed. The draftsman will have to give whatever information is necessary if any of these operations are required.

The mold will also require the attention of the draftsman. The designer of the mold will provide much of the information that the draftsman will be required to include in the drawing. The draftsman, however, should become familiar with the terms in use and their meanings.

WORDS AND PHRASES USED BY THE PLASTIC DRAFTSMAN

Wall thickness for compression molds should not be less than 1/16 of an inch. It may be more, but the thicker wall increases molding time. The wall should, therefore, be as thin as possible. For injection molds, walls should not be less than .050 inches thick, unless high pressure is used, and the plastic hasn't far to go.

As Fig. 398 illustrates, the wall should be of even thickness throughout the mold and should be thick enough to withstand the pressure required to eject the finished object from the mold.

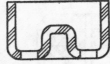 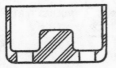

UNIFORM THICKNESS   VARYING THICKNESS

PREFERRED   NOT RECOMMENDED

Fig. 398.

Ribs, as shown in Figs. 399 and 401, are used to strengthen large unsupported surfaces and to make it easier for the plastic to flow into a mold by serving as runners. Ribs should taper 2 to 5 degrees, be rounded at the base and on the top. The rib, at its base, should be one-half the thickness of the base on which it rests.

Sometimes two ribs make the plastic flow easier.

To avoid sinking, when a rib is used, small ribs are put above the rib to compensate for the area that will sink. This is shown in Fig. 400.

Bosses may be used for mounting or for reinforcing holes. In order to eliminate the necessity of a finishing operation, three bosses are considered more advantageous than four. See Fig. 402.

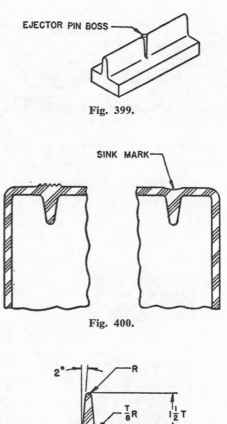

EJECTOR PIN BOSS

Fig. 399.

SINK MARK

Fig. 400.

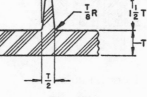

Fig. 401.

Draft means a slight spreading out of a mold to allow the object within to come out more easily. Fig. 403 illustrates this. A draft may vary from a quarter of a degree to four degrees per side, depending upon the material used and the way the finished object comes out of the mold. The draftsman must give whatever draft information is required. Sometimes no draft is necessary. In close quarters a maximum or minimum dimension may be needed, as shown in Fig. 404.

Fig. 402.

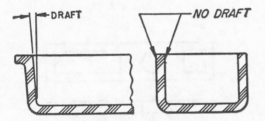

■■° DRAFT PER SIDE UNLESS OTHERWISE SPECIFIED.

Fig. 403.

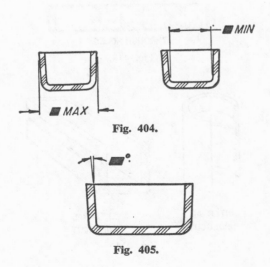

Fig. 404.

Fig. 405.

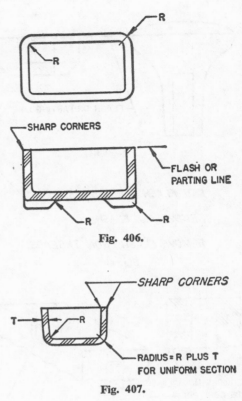

Fig. 406.

Fig. 407.

Taper looks like a draft, but is an angle that is part of the design of the object. Notice the taper in Fig. 405.

Fillets and radii are small curves in corners. They make the plastic flow smoother and ejection of the finished object easier. Fillets and radii remove dust traps and lines that may result from the flow of the plastic.

The radius of all corners should be one-half the thickness of the wall, but not less than 1/32 of an inch at any time except for outside corners. See Fig. 406. If the corner edge is to be sharp, the draftsman should note it, as shown in Fig. 406. He should also give the radius of the fillets and the outside radius, if the wall of the section must be the same thickness throughout. See Fig. 407.

If several corners have radii, the draftsman should indicate it in a general note.

Flash Lines and Gates. When objects are mass produced, they usually connect at one point. This is the point at which the objects are separated. The small amount of material left on

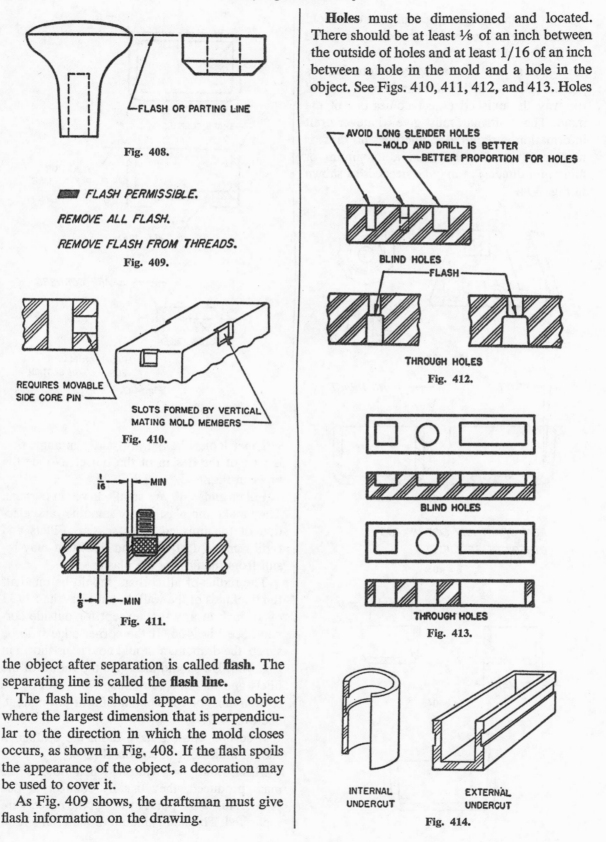

FLASH OR PARTING LINE

Fig. 408.

FLASH PERMISSIBLE.

REMOVE ALL FLASH.

REMOVE FLASH FROM THREADS.

Fig. 409.

REQUIRES MOVABLE
SIDE CORE PIN

SLOTS FORMED BY VERTICAL
MATING MOLD MEMBERS

Fig. 410.

Fig. 411.

Holes must be dimensioned and located. There should be at least ⅛ of an inch between the outside of holes and at least 1/16 of an inch between a hole in the mold and a hole in the object. See Figs. 410, 411, 412, and 413. Holes

AVOID LONG SLENDER HOLES
MOLD AND DRILL IS BETTER
BETTER PROPORTION FOR HOLES

BLIND HOLES

FLASH

THROUGH HOLES

Fig. 412.

BLIND HOLES

THROUGH HOLES

Fig. 413.

INTERNAL
UNDERCUT

EXTERNAL
UNDERCUT

Fig. 414.

the object after separation is called **flash.** The separating line is called the **flash line.**

The flash line should appear on the object where the largest dimension that is perpendicular to the direction in which the mold closes occurs, as shown in Fig. 408. If the flash spoils the appearance of the object, a decoration may be used to cover it.

As Fig. 409 shows, the draftsman must give flash information on the drawing.

should be as large as possible. Avoid long, thin holes. Whenever possible, use through holes instead of blind holes.

Undercuts, shown in Fig. 414, prevent the object from being easily removed from the mold.

Inserts are pieces or parts that are placed in the plastic material before forming so that they will be part of the finished object. Fig. 415 shows inserts loaded in the same half of the mold and inserts loaded in both halves of the mold. Long, slender inserts tend to break or bend unless properly supported. Both ends of such inserts should be supported. Inserts should not be too close to a wall, should be chamfered at the end that sticks out, should be deep enough in the object, and the part of the insert in the object should have a rough or knurled surface. See Figs. 416, 417, 418, 419, 420, 421, and 426. Fig. 421 shows that clearance is necessary

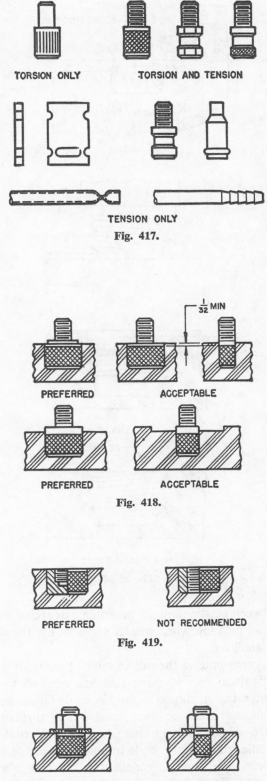

Fig. 417.

Fig. 418.

Fig. 419.

Fig. 420.

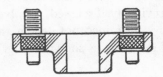

INSERTS LOADED IN
SAME HALF OF MOLD

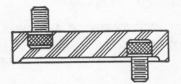

REQUIRES INSERTS TO BE LOADED
IN BOTH HALVES OF MOLD

Fig. 415.

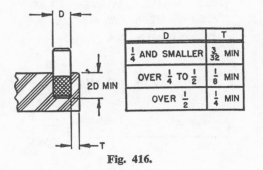

D	T
$\frac{1}{4}$ AND SMALLER	$\frac{3}{32}$ MIN
OVER $\frac{1}{4}$ TO $\frac{1}{2}$	$\frac{1}{8}$ MIN
OVER $\frac{1}{2}$	$\frac{1}{4}$ MIN

Fig. 416.

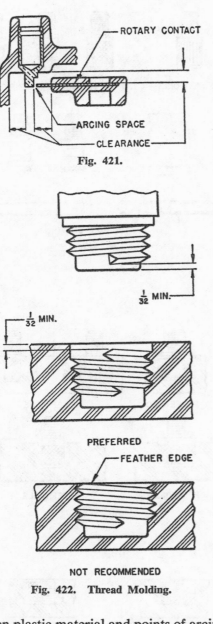

Fig. 421.

Fig. 422. Thread Molding.

between plastic material and points of arcing to avoid burning when inserts serve as electrical conductors.

Avoid putting threads in molds because it is difficult to remove parts that are screwed together. Inserts should be used instead of threads. If threads must be used, coarse-pitch threads 5/16 of an inch in diameter are preferred. Smaller threads are made in holes that are first molded and then threaded after the object is formed. For an illustration of thread molding, see Fig. 422.

As shown in Figs. 423 and 424, the draftsman must note the surface quality of objects. He must include the smoothness, gloss, or color and indicate where these surface finishes may be affected by temperature or chemicals.

Warpage may occur in large areas of plastic materials. The draftsman should note the permissible amount of warpage, as shown in Fig. 425.

Tolerances are difficult to maintain in plastic molding. Therefore, very close tolerances are seldom requested. The amount of tolerance depends on the kind of plastic, variation in batches of the same material, molding technique, size and shape of the part, and location of the parting line.

ASSEMBLY METHODS

Parts are very often molded individually and then assembled to form the desired object. One method, previously discussed, involves the use of metallic inserts incorporated in the part at the time of molding, as shown in Fig. 415. Fig.

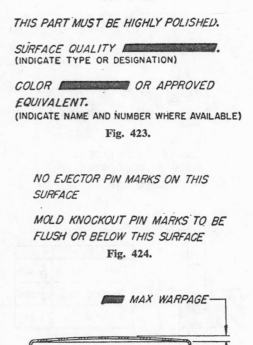

THIS PART MUST BE HIGHLY POLISHED.

SURFACE QUALITY ▬▬▬.
(INDICATE TYPE OR DESIGNATION)

COLOR ▬▬▬ *OR APPROVED EQUIVALENT.*
(INDICATE NAME AND NUMBER WHERE AVAILABLE)

Fig. 423.

NO EJECTOR PIN MARKS ON THIS SURFACE

MOLD KNOCKOUT PIN MARKS TO BE FLUSH OR BELOW THIS SURFACE

Fig. 424.

Fig. 425.

426A shows that inserts, including thermoplastics when heat is not a factor, can also be used to improve the appearance of the object. Other molded-in inserts fasten parts together, as shown in Fig. 426B, and serve as part of the electrical circuit, as shown in Fig. 426C.

The design of these molded-in inserts should be drawn on a separate detail drawing showing all parts large and clear, with full dimensions and notes.

Parts can also be fastened together by molding matching holes in each part. Rivets, screws, bolts, eyelets, or studs can be inserted in these matching holes after molding. See Fig. 427.

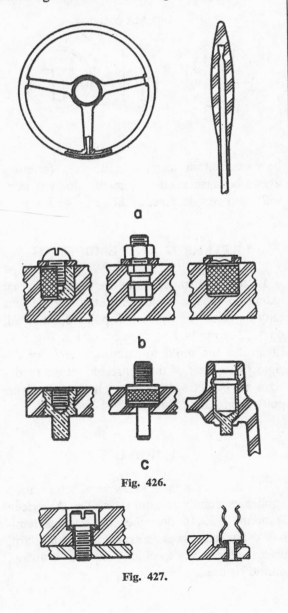

Fig. 426.

Fig. 427.

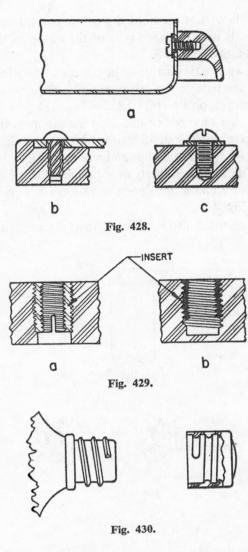

Fig. 428.

Fig. 429.

Fig. 430.

Another method of fastening includes the use of screws. When parts do not have to be disassembled often, self-tapping screws are used. Plain holes with the proper diameter are molded and the screws are inserted later. See Fig. 428A. For permanent assemblies, drive screws, shown in Fig. 428B, 1/16 of an inch to 5/16 of an inch in diameter are used. Machine screws, shown in Fig. 428C, can also be used to fasten parts.

Separate threaded inserts, shown in Fig. 429, are sometimes used for fastenings. Fig. 429A shows a self-tapping insert with a threaded hole. Fig. 429B shows a wire coiled to match the thread size before it is screwed into an already tapped hole in the plastic.

When parts are molded, internal and external threads may be made part of the shape of the mold. See Fig. 430.

Two plastic parts may be fastened by shrinking one part on the other, or a plastic part may be shrunk on a metal part. See Fig. 431.

Other methods of fastening include pressing in metal parts, Fig. 432; spring hinges and clips, Fig. 433; and stud clips and retainers, Fig. 434.

Thermoplastic parts may be joined by using pressure and a hot punch to re-form one part. See Fig. 435.

Cement is also used for fastening. Care must

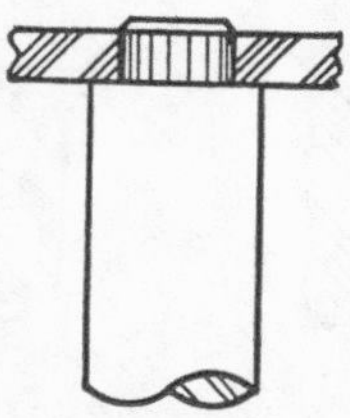

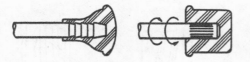

Fig. 431.

Fig. 432.

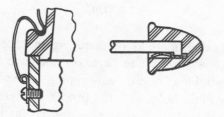

Fig. 433.

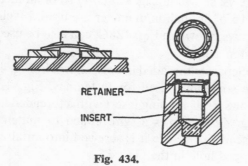

RETAINER

INSERT

Fig. 434.

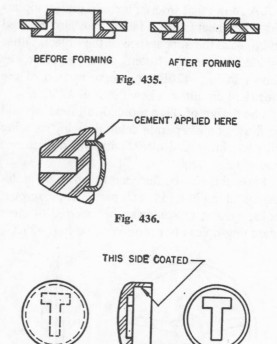

BEFORE FORMING   AFTER FORMING

Fig. 435.

CEMENT APPLIED HERE

Fig. 436.

THIS SIDE COATED

Fig. 437.

be exercised in using cement as a fastening agent because certain cements do not work well with certain plastics. See Fig. 436.

LETTERING AND EMBOSSING

Letters and trade-marks are used to a great extent in plastic molding. Letters or designs may be raised or depressed, painted, stamped, coated, or applied as decalcomanias. Fig. 437 illustrates the mold for coating letters or designs. The back of the depressed letters or designs are coated with copper, chromium, silver, gold, bronze, or paint.

LAMINATES

The draftsman must know what information applies especially to laminates. The American Standard C59.16 describes the various laminates. The tolerances described in this standard, unless otherwise specified on the drawing, should be used.

The amount of lamination that should be used depends upon the load that will be applied to the finished object. Fig. 438 shows the possible types of lamination arrangement for different types of loads. Fig. 439 shows the recommended spacing between holes. As large a hole as possible should be used in cantilevered laminates. See Fig. 440.

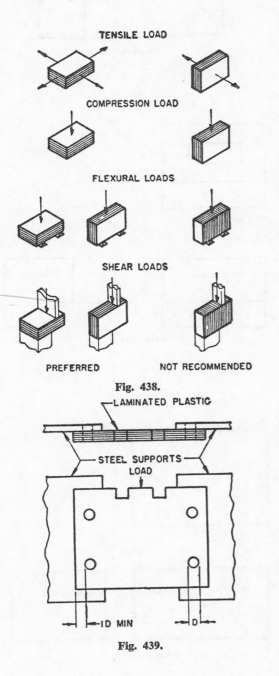

TENSILE LOAD

COMPRESSION LOAD

FLEXURAL LOADS

SHEAR LOADS

PREFERRED   NOT RECOMMENDED

Fig. 438.

LAMINATED PLASTIC

STEEL SUPPORTS
LOAD

ID MIN   D

Fig. 439.

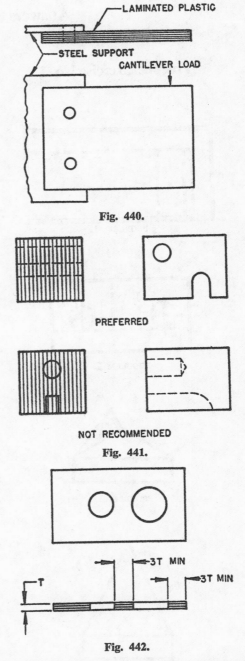

LAMINATED PLASTIC

STEEL SUPPORT
CANTILEVER LOAD

Fig. 440.

PREFERRED

NOT RECOMMENDED

Fig. 441.

3T MIN
T   3T MIN

Fig. 442.

As Fig. 441 shows, all drawing and machining should be done at right angles to the laminations.

Laminated materials can be punched up to a thickness of ⅛ of an inch. Paper-based laminates can be sheared up to a thickness of 1/16 of an inch. Most fabric-based laminates can be sheared up to a thickness of ⅛ of an inch. The holes should be placed as shown in Fig. 442.

Answers to Practice Exercises

Practice Exercise No. 1

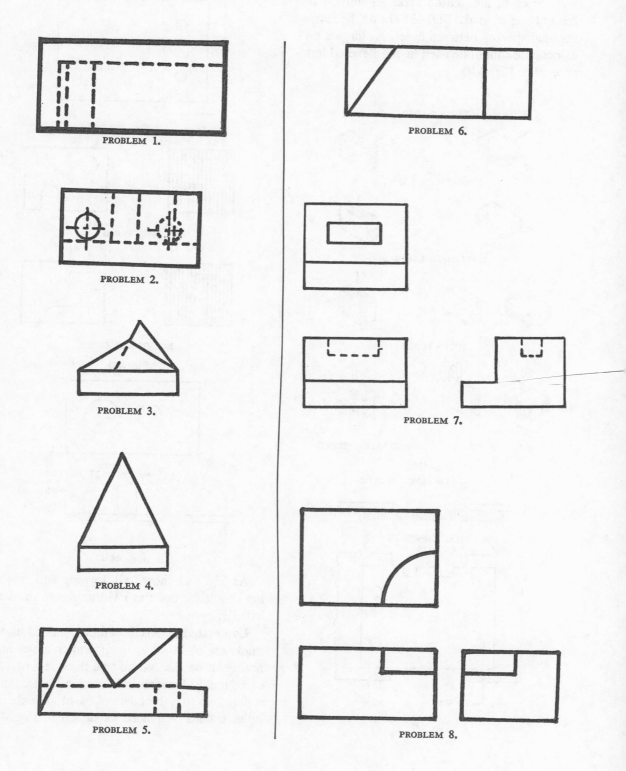

PROBLEM 1.

PROBLEM 2.

PROBLEM 3.

PROBLEM 4.

PROBLEM 5.

PROBLEM 6.

PROBLEM 7.

PROBLEM 8.

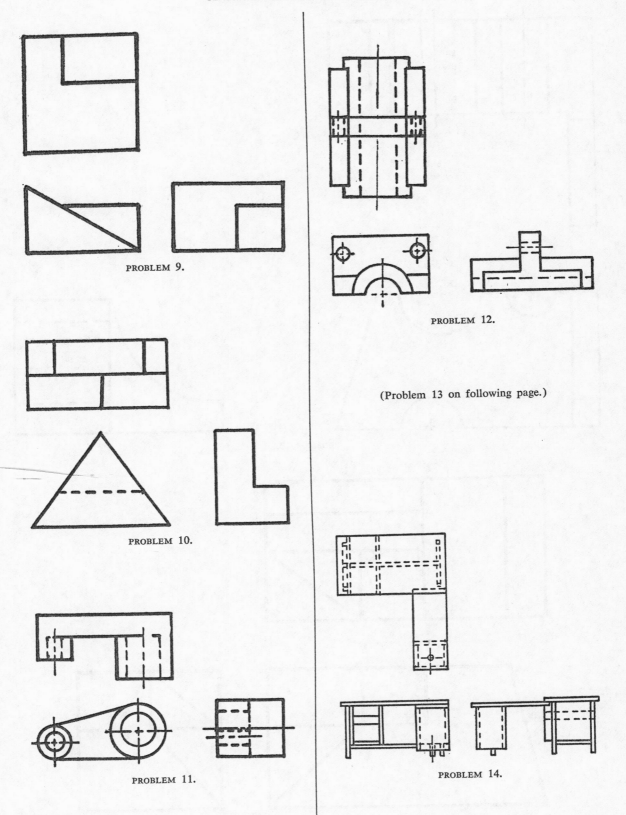

PROBLEM 9.

PROBLEM 10.

PROBLEM 11.

PROBLEM 12.

(Problem 13 on following page.)

PROBLEM 14.

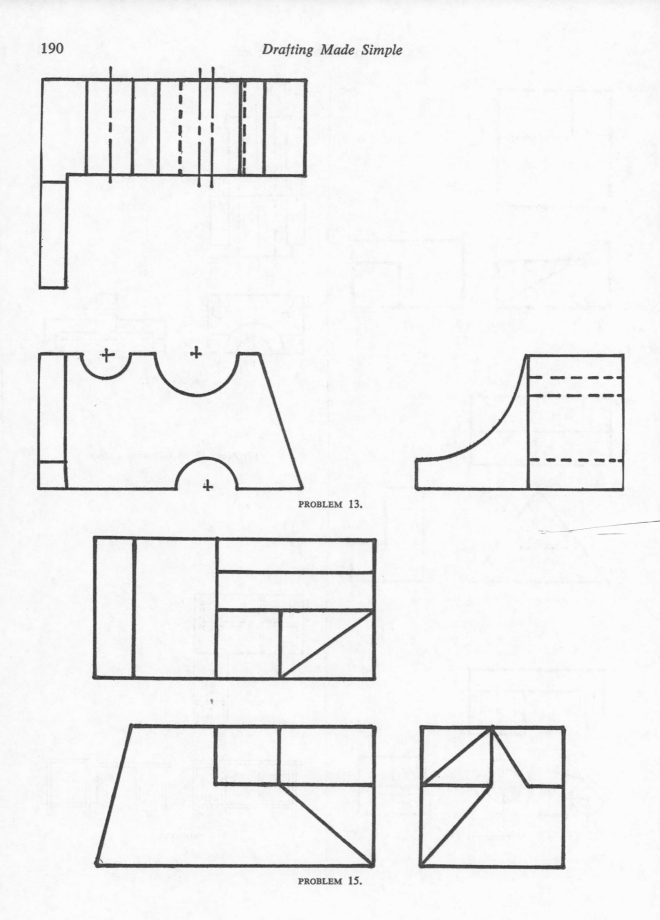

PROBLEM 13.

PROBLEM 15.

Practice Exercise No. 2

SECTION A-A

SECTION B-B

SECTION C-C

PROBLEM 1.

Practice Exercise No. 3

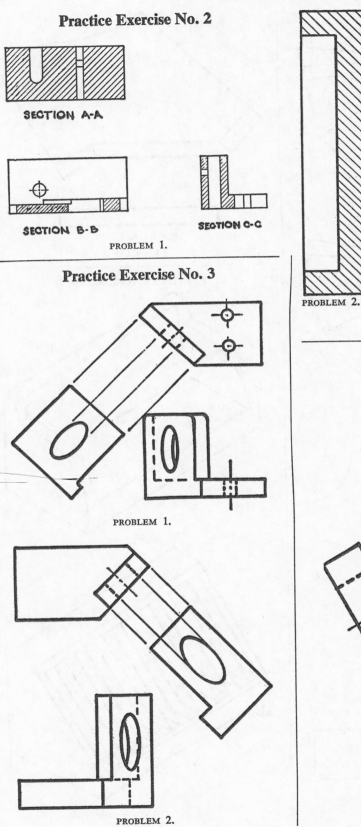

PROBLEM 1.

PROBLEM 2.

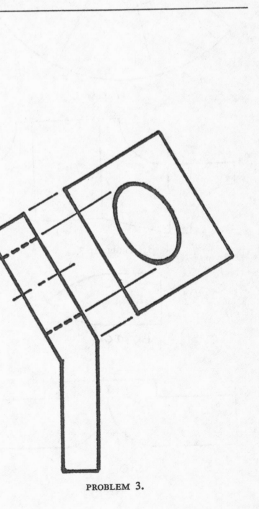

PROBLEM 2.

PROBLEM 3.

PROBLEM 3.

Practice Exercise No. 4

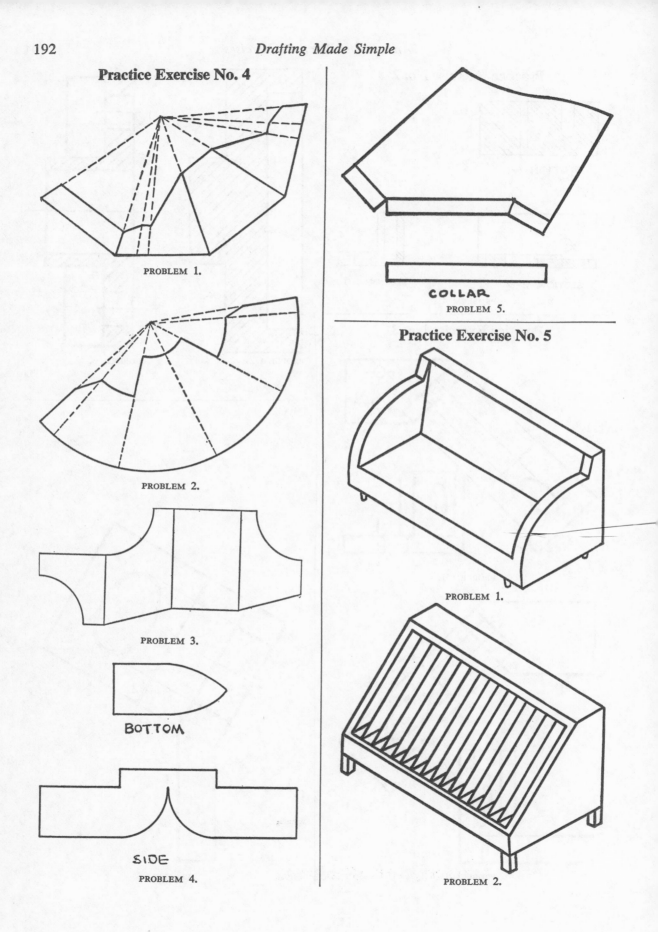

PROBLEM 1.

PROBLEM 2.

PROBLEM 3.

BOTTOM

SIDE

PROBLEM 4.

COLLAR

PROBLEM 5.

Practice Exercise No. 5

PROBLEM 1.

PROBLEM 2.